EINSTEIN'S UNIFICATION

Why did Einstein vainly study unified field theory for more than 30 years? In this book, the author argues that Einstein believed he could find a unified theory of all of nature's forces by repeating the methods he is thought to have used when he formulated general relativity.

The book discusses Einstein's route to the general theory of relativity, focusing on the philosophical lessons that he learnt. It then addresses his quest for a unified theory for electromagnetism and gravity, discussing in detail his efforts with Kaluza–Klein theory and, surprisingly, the theory of spinors. From these perspectives, Einstein's critical stance towards quantum theory comes to stand in a new light. This book will be of interest to physicists, historians and philosophers of science.

JEROEN VAN DONGEN is Assistant Professor at the Institute for History and Foundations of Science at Utrecht University, the Netherlands, and has served as Editor of the *Collected Papers of Albert Einstein* at the California Institute of Technology.

EINSTEIN'S UNIFICATION

JEROEN VAN DONGEN

Utrecht University
and
Einstein Papers Project,
California Institute of Technology

CAMBRIDGE
UNIVERSITY PRESS

CAMBRIDGE
UNIVERSITY PRESS

University Printing House, Cambridge CB2 8BS, United Kingdom

One Liberty Plaza, 20th Floor, New York, NY 10006, USA

477 Williamstown Road, Port Melbourne, VIC 3207, Australia

314-321, 3rd Floor, Plot 3, Splendor Forum, Jasola District Centre, New Delhi - 110025, India

79 Anson Road, #06-04/06, Singapore 079906

Cambridge University Press is part of the University of Cambridge.

It furthers the University's mission by disseminating knowledge in the pursuit of education, learning and research at the highest international levels of excellence.

www.cambridge.org
Information on this title: www.cambridge.org/9781108703031

© J. van Dongen 2010

This publication is in copyright. Subject to statutory exception and to the provisions of relevant collective licensing agreements, no reproduction of any part may take place without the written permission of Cambridge University Press.

First published 2010
Reprinted 2011
First paperback edition 2018

A catalogue record for this publication is available from the British Library

ISBN 978-0-521-88346-7 Hardback
ISBN 978-1-108-70303-1 Paperback

Cambridge University Press has no responsibility for the persistence or accuracy of URLs for external or third-party internet websites referred to in this publication, and does not guarantee that any content on such websites is, or will remain, accurate or appropriate.

Contents

Acknowledgments

This book began as a thesis project, with the loosely prescribed perimeter of Albert Einstein's unified field theory work as a subject. The advisors that dedicated themselves to my project could not have been chosen better: historian of modern physics Anne J. Kox and string theorist Herman Verlinde. Anne taught me many things, Einstein and otherwise, and has been extremely supportive of, and important to my work. I am proud and very grateful that he has continued his advice and friendship well after my days as a graduate student ended; between us a bond has formed that easily surpasses that of professor and student. Similarly, this book would not have come into being without Herman's manifold contributions. I have fond recollections of the many engaging discussions and blackboard instructions, and Herman, too, has given much more than can be captured by the title of thesis advisor alone; I have greatly valued his intellectual stewardship, friendship and extensive hospitality in New Jersey.

The project started at Anne Kox's history of physics group at the Institute for Theoretical Physics of the University of Amsterdam. The Institute's faculty, postdoctorate and student members provided the best intellectual environment any young scholar can wish for: applying a high standard while maintaining a warm atmosphere. I am honored to be a graduate of the University of Amsterdam. I also greatly enjoyed the hospitality that was awarded me as a visitor for nearly two years at Princeton University's Physics Department, by both its faculty and most welcoming fellow graduate students.

Starting from a loosely prescribed subject, my work took a major turn when I learnt of the work conducted on the history of general relativity by Michel Janssen, John Norton, Jürgen Renn, Tilman Sauer and John Stachel. Only when I became acquainted with their studies did I begin to see the outline of my thesis project, and ultimately this book. I have further heavily drawn upon the work of such eminent Einstein scholars as Gerald Holton, Abraham Pais, Vladimir Vizgin, and many more.

How does this book differ from my earlier thesis? Postdoctoral scholarships and editorial responsibilities have made me much more seasoned in Einstein scholarship, and history of science in general, and this has expanded the book in proportion. Following my graduation in 2002 I had the privilege of becoming a member of Jürgen Renn's group at the Max Planck Institue for the History of Science in Berlin, until 2004, when I joined the Einstein Papers Project, directed by Diana Kormos Buchwald at the California Institute of Technology. Both experiences have given me a deepened understanding of Einstein's relation to his science and a sharpened sense for historical scholarship. This book benefitted greatly from these experiences. Interactions in Berlin with many of the Institute's permanent and transient members – Jürgen Renn, Michel Janssen, Dieter Hoffmann, Giuseppe Castagnetti and Matthias Schemmel must be mentioned in particular – gave me a fuller sense of Einstein's road to relativity, as well as of many other aspects of his science. At the Einstein Papers Project, my maturing as a historian of science continued, and in Pasadena I have enjoyed strong support and most valuable exchanges of ideas with Diana Buchwald, Tilman Sauer, József Illy, Ze'ev Rosenkranz and Dan Kennefick; I am also grateful for the help I received from Carol Chaplin, Rudy Hirschmann, Rosy Meiron, Osik Moses and Jennifer Nollar. In particular my views on Einstein's relation to quantum mechanics were formed while studying his correspondence for our edited volumes; the reconstruction of Einstein's life in day-to-day detail provides enthralling insight into his science. It is a privilege to serve as a member of the Einstein Papers team, and I wish to thank Diana for her persistent help in bringing this book to a successful close.

Lately I have found a true home at the Institute for History and Foundations of Science at Utrecht University, where I have had the opportunity to join the excellent group headed by Dennis Dieks. Dennis has been most helpful in steering this book project into final form by serving as a critical reader and dedicated sounding board; he has given me thoughtful and much valued advice on all of its subjects. I have further greatly appreciated his careful coaching of my teaching; I am indebted to Dennis for any future success I may enjoy as a lecturer. In Utrecht I have further found a dedicated colleague in Jos Uffink, who also has given me support and direction in completing this project. At the Institute, I have aspired to act as someone resembling an intellectual trait d'union between the foundationalists and historians, and I thank my historian colleagues, Bert Theunissen, Lodewijk Palm and honorary member Frans van Lunteren for their encouragement. I also owe gratitude to Utrecht's many smart graduate and undergraduate students who by their probing questions and dedicated work make doing research and teaching in Utrecht a most joyful experience. Finally, Utrecht has recently created a unique infrastructure for history and philosophy of science with the Descartes Center; in it, both humanities and science faculties have been unified, and I look forward to fulfilling my role as one of its members.

In preparation for this study I have consulted Einstein's personal correspondence, as contained in duplicates of the Einstein Archive, an invaluable research tool, the original of which is to be found at the Hebrew University of Jerusalem, as per Einstein's last will. I depended on the copy of Princeton University's Firestone library and, more recently, on another, updated copy in the basement of the Einstein Papers Project in Pasadena. This resource has been most helpful; in fact, without its many revealing documents it would have been well-nigh impossible to formulate my ideas. I thank Roni Grosz of the Albert Einstein Archives and Daphne Ireland at Princeton University Press for permission to use the Archive's material in this book. Please note that

• references in the footnotes to the Einstein Archive will be indicated with the abbreviation EA.

An earlier version of Chapter 5 on semivectors has appeared in the *Archive for History of Exact Sciences*, and I thank its publisher, Springer, for permission to use the material in this book.

This book has benefitted from many proofreaders. For their engaging criticism, I would like to express my sincere gratitude to Eddy Ardonne, Sander Bais, Michel Janssen, Jürg Käppeli, Dan Kennefick, Klaas Landsman, Ad Maas, John Norton, Tilman Sauer, Michiel Seevinck, Jürgen Renn, Ze'ev Rosenkranz and Jos Uffink. I am grateful to Rob van Gent for help with figures, and Albert van Helden for advice. I have benefitted greatly from interacting with many historians and philosophers of science of my generation during my *Wanderjahre* as postdoc; I would like to thank in particular Sven Dupré, Christoph Lehner, Alberto Martinez, Suman Seth, Chris Smeenk, Milena Wazeck and Daan Wegener for sharing their insight and company with me. I thank John Norton for much appreciated discussions, advice and help. Discussions with Martin J. Klein inspired me to write Chapter 4 on experiment. Martin and his wife Joan Blewett invited me for a lovely week at their home in North Carolina in 2005, and on that occasion we committed ourselves to ambitious collaborative plans, yet busy schedules regrettably inhibited progress. I was greatly saddened, and it was a great loss to the scholarly community, when Martin passed away in March of 2009.

I thank the Netherlands Organization for Scientific Research (NWO) for supporting me with a generous Veni grant, and for support during my graduate study. I am further most grateful to the Pieter Zeeman Foundation and the Center for High Energy Astrophysics of the University of Amsterdam for their financial aid. I am honored that the Society for the Advancement of Science, Medicine, and Surgery,

Amsterdam, chose to award my thesis its Andreas Bonn medal for 1999–2003. At Cambridge University Press, I wish to thank Simon Capelin, whose enthusiasm at our first contact was very motivating, Laura Clark, Lindsay Barnes and Graham Hart, and Sehar Tahir of CUP's tex support.

I wish to thank my friends for their patience and help, scattered as they are from the heart of Brabant to the Pacific coast of the American continent. My father deserved to see this book appear; sadly, he passed away too early. This book could never have been written without the unfailing support of my loving and dedicated mother, Mia Savenije. I thank you, Rocio, for your love.

Utrecht, July 2009

Introduction

"This exposition has fulfilled its purpose when it shows the reader how a life's efforts are related to one another and why they have led to expectations of a particular kind."[1] Thus end Albert Einstein's "Autobiographical notes," as we begin our account. Einstein's critical self-assessment was published in a collection of essays entitled *Albert Einstein: Philosopher-Scientist*, and it appeared in 1949, only six years before Einstein's final passage. These "Notes" constitute a rich source of personal reflections on his life and science; in particular, they discuss in detail the ideas that dominated his later years. The latter are the subject of this book, and it is with the above sense of purpose that we wish to approach it.

Thus, this study will be about Einstein's search for a unified theory for all physical forces, and its relation to his larger oeuvre in science and philosophy. Our story starts with the events leading up to the 1915 discovery of general relativity and its scope extends to Einstein's passing away in 1955. We have a vast subject, and it will be impossible to address minutely all its elements; both some familiar and some less familiar themes of Einstein's later physics will have to remain undiscussed. What we will present, rather, is an attempt, from a historical perspective, at a synthesis, and our selection will be aimed to serve that synthesis.

When leaving out of consideration the elaborations of general relativity, one can characterize Einstein's later work according to roughly two strands: on the one hand there are his numerous publications on classical unified field theories, and on the other hand we find his critique of quantum theory. The secondary literature on the latter subject is quite voluminous. Yet, the larger part of this literature only superficially engages with Einstein's efforts in unified field theory, despite the fact that

---

[1] "Diese Darlegung hat ihren Zweck erfüllt, wenn sie dem Leser zeig[t], wie die Bemühungen eines Lebens miteinander zusammenhängen und warum sie zu Erwartungen bestimmter Art geführt haben." As in Einstein (1949a), p. 94.

he published about twice as many papers on this topic as on quantum mechanics.[2] This is a rather unfortunate oversight, in particular when one realizes that these efforts were intended to yield an alternative for the quantum theory – that they did not bear any fruit can hardly justify the lacuna. One of our aims will be to see how and why Einstein chose to pursue unified field theory on the one hand, and on the other hand how this related to his dissenting attitude to quantum theory. Thus we hope to arrive at a coherent perspective on Einstein's later efforts.

A central observation in our discussion will be that Einstein's increasing engagement with unified field theory and his distance from the quantum program developed in tandem with a changing recollection of his route to the general theory of relativity. From the perspective of the older Einstein, it seemed as if the successful formulation of the latter theory had been the result of seeking the mathematically most natural formulation for a generally relativistic gravity theory. This recollection furthermore became a point of reference, even of justification, for his later work. These developments were well reflected in Einstein's changing views on the method of science, just as these views influenced as well as mirrored his contemporary practice in theory construction. The following excerpt from a letter that Einstein sent to Louis de Broglie in 1954 (just when David Bohm sought to revive de Broglie's ideas about hidden variables) illustrates our perspective rather well:

That I am writing you has a peculiar reason. Namely, I would like to tell you *how* I have come to my method, which from the outside must seem quite bizarre: I must look like an ostrich that keeps his head buried in relativistic sand so that he does not have to look the evil quanta in the eye. In reality I am, just like you, convinced that one should look for a substructure, the necessity of which has been cleverly disguised by the current quantum theory through its use of the statistical mold.

I have however long been convinced that one shall not be able to find this substructure *in a constructive way* from the known empirical relations between physical things, because the required mental leap would exceed human powers. I have arrived at this opinion not only because of the fruitlessness of the efforts of many years, but rather also through the experiences with the gravitation theory. The gravitational equations could *only* be found by a purely formal principle (general covariance), that is, by trusting in the largest imaginable logical simplicity of the natural laws. As it was obvious that the theory of gravity constitutes only a first step in finding the simplest possible general field laws, it seemed to me that this logical route should first be thought through to the end before one can hope to arrive also at

---

[2] An insightful exception to the above mentioned literature is Abraham Pais's biography (1982), see in particular p. 328, pp. 460–469.

a solution of the quantum problem. This is how I became a fanatic believer in the method of "logical simplicity." [...] This should explain the ostrich policy.[3]

Einstein acknowledged that, in the eyes of his contemporaries, he must have looked like an ostrich in his obstinate attitude to quantum theory. But he felt that his actual position was more subtle: he had not buried his head in the sand, but was rather trying to look beyond the statistical quantum theory. In this attempt, he cited as his foremost guiding principle a maxim of "logical simplicity" – we will see that we can just as well read here "mathematical naturalness." Both were intimately related to his quest for unification, and to understand better the motivations of this striving, as the letter to de Broglie indicates, we have to involve his struggles to find the general theory of relativity.

After the formulation of general relativity Einstein began playing down the importance of induction from the phenomena. He then added to this an emphasis on the creative merit of aiming for mathematical simplicity and naturalness. When these epistemological ideas had settled, they found their clearest public expression in his 1933 Herbert Spencer lecture at the University of Oxford, "On the method of theoretical physics." The contents of that lecture will be an important element in our story. Its ideas were later, in 1952, concisely captured in a schema that Einstein drew in one of his letters to his friend Maurice Solovine, and this schema will prove to be quite practical in describing Einstein's intellectual development.

When writing to de Broglie, Einstein indeed stated that he believed in a particular "method" while he was engaged in his latest efforts. In the second half of his professional career, his view "on the method of theoretical physics" became the premier vantage point from which he would assess his involvement with his discipline, and we will pick up on these references. In recent historiography of science, however, methodological perspectives have regularly been discounted as indicative of an outdated approach to history; in this outdated approach, history was supposedly regarded as the hand-maiden of philosophy, and barely rose above presenting a

---

[3] "Dass ich Ihnen nun schreibe, hat eine eigentümliche Ursache. Ich will Ihnen nämlich sagen, *wie* ich zu meiner Methodik getrieben worden bin, die von aussen gesehen recht bizarr ist. Ich muss nämlich erscheinen wie der Wüsten-Vogel Strauss, der seinen Kopf dauernd in dem relativistischen Sand verbirgt, damit er den bösen Quanten nicht ins Auge sehen muss. In Wahrheit bin ich genau wie Sie davon überzeugt, dass man nach einer Substruktur suchen muss, welche Notwendigkeit die jetzige Quantentheorie durch Anwendung der statistischen Form kunstvoll verbirgt. Ich bin aber schon lange der Überzeugung, dass man diese Substruktur nicht *auf konstruktivem Wege* aus dem bekannten empirischen Verhalten der physikalischen Dinge wird finden können, weil der nötige Gedankensprung zu gross wäre für die menschlichen Kräfte. Zu dieser Meinung kam ich nicht nur durch die Vergeblichkeit vieljähriger Bemühungen, sondern auch durch die Erfahrungen bei der Gravitationstheorie. Die Gravitationsgleichungen waren *nur* auffindbar auf Grund eines rein formalen Prinzips (allgemeine Kovarianz), d.h. auf Grund des Vertrauens auf die denkbar grösste logische Einfachheit der Naturgesetze. Da es klar war, dass die Gravitationstheorie nur einen ersten Schritt zur Auffindung möglichst einfacher allgemeiner Feldgesetze darstellt, schien es mir, dass dieser logische Weg erst zu Ende gedacht werden muss, bevor man hoffen kann zu einer Lösung auch des Quantenproblems zu gelangen. So wurde ich zu einem fanatischen Gläubigen der Methode der „logischen Einfachheit". [...] Dies zur Erklärung der Vogel-Strauss-Politik." Einstein to Louis de Broglie, 8 February 1954, EA (i.e., Albert Einstein Archives, Jerusalem) 8-311. Emphasis as in original.

cumulative account that suggested an inevitable progress in knowledge, effected by contributions of great geniuses. As philosophy would try to capture the "right" method, reflective of a timeless human *ratio*, history's task was foremost to provide case studies that either illustrated or disproved.[4]

This approach has been replaced by a historiography that regards physics, and other scientific disciplines, as part of the larger culture, and rightfully so – even if its excesses did for a while turn history of science into the battleground of an unfortunate culture war that pitted science against the humanities. The new perspective has given the opportunity to answer a multitude of insightful questions, some of which, in the old viewpoint, could hardly even have been raised. Thus, for example, we now have insight into how physics as an academic discipline – essentially a cultural niche – was formed out of a variety of scientific groups and practices. Yet, to discount wholly the issue of method has overshot the mark. In the particular case of Einstein, a discussion of methodology might in fact be unavoidable, as he expressed himself with regular reference to his "method," not unlike other physicists of his generation. As we will argue, his methodology furthermore developed in interaction with both his current and past work in physics; understanding Einstein as a practicing scientist therefore inevitably leads one to consider his methodological stances.

Our discussion of methodology should be viewed in that light: we do not aspire to capture "Einstein's method" in order subsequently to chastise its failures or instruct a chase after its successes. Rather, we intend to approach methodology as an inalienable, active element of Einstein's engagement with his science, just as it was a reflection thereof. After all, it was Einstein himself who, particularly in his later years, deliberated on his practice in theoretical physics at a methodological meta-level. This circumstance should be of benefit to historians as it gives crucial insight into how he expected theoretical physics ought to describe and explain the natural phenomena – how theoretical physics should and should not be done. The methodological perspective is thus a valuable instrument when trying to put Einstein's historical contingency in a sharper light.

From the perspective of the Spencer lecture "on the method," we will first look back on the formulation of general relativity, with a particular focus on the heuristic strategies that Einstein employed. The frustrations and successes of these strategies shed light on the gradual epistemological reorientation that Einstein subsequently underwent. The ideas of the Spencer lecture were to Einstein a personal reflection on his own work, rather than a chiselled charter for theoretical research. We will see, however, that he also engaged his methodological ideas, and the discovery of general relativity, to find his direction in his later efforts in unified field theory. The

---

[4] For this characterization of the outdated approach, see for example Morus (2005), pp. 4–5.

subject in which we will study this interaction in particular is his research on the "semivector," the topic he was pursuing at the time of his lecture in Oxford.

To come to grips with Einstein's ideal for a physical theory, we give an outline of his encounter with the Kaluza–Klein theory. As for any of the unified field theories that he studied, his hope was that this theory could somehow undercut the quantum theory. This hope was partly motivated by his ideas on how a physical theory should give explanations of the phenomena – how a theory could yield understanding of nature. As suggested, these ideas were closely related to his methodological convictions.

In the second paragraph of the letter to de Broglie quoted above, Einstein criticized in an implicit way the heuristics of the quantum theorists. These were, much more than his own, motivated by an empirically oriented philosophy – that was in fact rather akin to the philosophical convictions he had held in his younger years. Such a methodology was to the later Einstein somewhat naive, and could only lead to a superficial description of nature.[5]

We will see that the difference in the practice of theory construction between Einstein and the quantum theorists was an important cause for his negative appraisal of quantum mechanics. This aspect complements in a natural way the objections that he addressed at the foundations of the theory; for example issues regarding completeness, determinism, etc.; the classical field theories under his scrutiny would furthermore presumably not exhibit these flaws. To Einstein, the foundational shortcomings of quantum theory were intimately linked to a conviction that experientially oriented methods could at best produce a mere phenomenological theory. Thus, as in the letter to de Broglie, Einstein pointed in particular to his alternative methodology to justify his rejection of quantum mechanics, and his pursuit of another theory.

---

[5] Einstein wrote to the mathematical physicist André Lichnerowicz, again in 1954: "If [...] the physical world cannot be reduced to logically simple elements, then there is no hope at all for us to understand things other than superficially," and continued by suggesting that both physicists and philosophers were currently too much under the sway of such superficial approaches. ("Wenn [...] die physikalische Welt nicht auf logisch einfache Elemente zurückführbar ist, dann gibt es für uns überhaupt keine Hoffnung, die Dinge anders als oberflächlich zu erfassen." Einstein to Lichnerowicz, January 1954, EA 16-319; for similar remarks, see also for example Einstein to Erwin Schrödinger, 22 December 1950, p. 36 in Einstein *et al.* (1963).)

1

Formulating the gravitational field equations

On 25 November 1915, Albert Einstein presented the final version of the field equations of the general theory of relativity to the Royal Prussian Academy of Sciences.[1] These equations were generally covariant: their form remained unchanged under arbitrary transformations of the space and time coordinates. This was a mathematical manifestation of Einstein's principle of equivalence, which held that the state of affairs in a homogeneous gravitational field is identical to the state of affairs in a uniformly accelerated coordinate system.

Einstein's first publication that contained the principle of equivalence appeared in 1907. It was included in a review paper of his relativistic account of electrodynamics of 1905. The principle immediately proved its heuristic value: on its basis Einstein already proposed in the same article the existence of a gravitational redshift of light, and the bending of light trajectories in a gravitational field.[2] Nevertheless, eight years would pass between the first formulation of the equivalence principle and its final vindication in 1915, when it acquired a firm footing in the field equations. During those years, Einstein remained nearly silent on gravitation from late 1907 until June 1911. He did not publish any substantial articles on the subject and, even more surprising, he hardly discussed it with his correspondents.

There are, however, strong indications that Einstein continued to think about the problems of gravity, as the equivalence principle and accelerated motions led to conceptual difficulties in the special theory of relativity. A discussion involving such luminaries as Max Laue and the mathematician Gustav Herglotz raged on the conflicts that rotation produced for straightforward definitions of relativistic rigid bodies. A rotating cylinder should have a Lorentz-contracting circumference, as seen from a reference system at rest, yet its radius – perpendicular to

---

[1] See Einstein (1915d).
[2] See Einstein (1907b); the above formulation of the principle of equivalence derives from Einstein's formulation on its p. 454; on Einstein's 1907 paper see also Miller (1992), Stachel (2007). For a historical discussion of the equivalence principle, see Norton (1985).

the rotation – would remain unaltered. Could one in this case still maintain the usual relation between radius and circumference from Euclidean geometry? The problem was known as the "Ehrenfest paradox," after Paul Ehrenfest, who had posed it in a concise one page paper in 1909.[3]

Einstein's silence ended in 1911, when he published an article that further elaborated the idea of gravitational light deflection and that included a first value for this effect in the case of light grazing the surface of the sun. Soon, Einstein set out to find field equations for a relativistic theory of gravity; first for the case of a static field, and then for the full dynamical case, spurred on by competing publications by Max Abraham and Gunnar Nordström.[4]

In the 1911 publication Einstein argued that the speed of light c should depend on the gravitational potential and thus on the spatial coordinates.[5] In his subsequent theory for the static field of 1912, the speed of light started playing the role of the potential itself. Einstein found that in a static field "local" time τ was to be related to "universal" time[6] t through $d\tau = cdt$. He pointed out that the equations of motion of a particle could easily be expressed in the Lagrangian formalism, in which they would take the form:

$$\delta \int \sqrt{c^2 dt^2 - dx^2 - dy^2 - dz^2} = 0. \tag{1.1}$$

Einstein observed that these equations "[let] us anticipate the structure of the equations of motion of a particle in a dynamical gravitational field."[7] Changing the coordinates would transform the single static gravitational potential c into ten functions $g_{\mu\nu}$, so Einstein may at this point have begun to surmise that the equation of motion for a particle coincides with the geodesic equation for non-flat spaces.[8]

The above equation suggests a departure from Euclidean geometry. This had already been suggested earlier by the rotating disk of the Ehrenfest paradox. Rotation involves inertial forces, and these can be seen as gravitational forces according to the equivalence principle. Thus, if in the case of a rotating disk one runs into trouble with Euclidean geometry, one could expect the same for the equivalent gravitational field.

---

[3] See Ehrenfest (1909); for historical discussion, see Stachel (1989a; 2007), Sauer (2007b).

[4] For the light reflection result, see Einstein (1911); Einstein's 1911 prediction was off by a factor of 2 when compared to the later theoretical and observed value; for a discussion, see Earman and Glymour (1980). Einstein's static theory is found in Einstein (1912a,b); for Abraham's and Nordström's theories, see Abraham (1912a), Nordström (1912), reprinted in English translation in Renn and Schemmel (2007a), pp. 331–339, 489–497.

[5] Einstein (1911), p. 906.

[6] Einstein (1912a), p. 366. In modern parlance, $d\tau$ would be a proper time increment and dt a time increment in coordinate time, in units in which a constant light velocity reduces to unity.

[7] "Die zuletzt hingeschriebene Hamiltonsche Gleichung lässt ahnen, wie die Bewegungsgleichungen des materiellen Punktes im dynamischen Gravitationsfelde gebaut sind," (Einstein 1912b), p. 458.

[8] See Stachel (2007), pp. 95–103.

In the summer of 1912, Einstein's former fellow student and friend Marcel Grossmann, now professor of mathematics at their alma mater, the *Eidgenössische Technische Hochschule* in Zurich, indicated that Einstein should take up the Riemannian theory of geometry and the relatively new mathematical theory of tensors.[9] Einstein wanted to find appropriate field equations for the $g_{\mu\nu}$-potentials. He soon realized, on the basis of elaborations of four-dimensional relativistic electrodynamics and hydrodynamics, that the ten-component stress-energy tensor was the right generalization of the Newtonian mass density in such field equations.[10] The central question then was, what term should the left-hand side of the gravitational field equations contain?

Recent scholarship – based on extensive studies of Einstein's actual research notes – has shown that Einstein was working along a two-pronged strategy as he tried to formulate candidate field equations.[11] On the one hand, he abstracted from a number of physical constraints. On the other hand, he looked for the laws that were most naturally written in the new tensor formalism. Einstein expected that both approaches would lead to the same results, but in 1913 it seemed to him that they would not. He had to choose, and he chose the first, the physical approach. This led to a first set of field equations that was, surprisingly, not generally covariant.

Einstein struggled to come to terms with his 1913 theory for two years. He recognized his errors towards the end of 1915, just as David Hilbert had turned his mathematical powers to the theory of gravity. Einstein spent November 1915 in strenuous labor, believing that Hilbert might be closing in on him. In this last month the natural mathematical constructions of Riemannian geometry and variational calculus pointed the way to the final field equations.

Nevertheless, Einstein's success should not be seen as exclusively the result of his newly acquired skills and sensibilities in mathematics, but rather as the result of a long learning process in which these skills and sensibilities had been interacting with a multifaceted complex of physical demands and beliefs. This learning process had brought him to the doorstep of the final theory, with the return to the Riemann tensor giving him the last push to the end result.

By 1933 Einstein was lecturing in Oxford that the creative principle in theory construction lies in mathematics, as by then he had become convinced that "nature is the realization of the simplest conceivable mathematical ideas."[12] He

[9] See Stachel (2007), pp. 101–108.

[10] Renn and Sauer (2007), pp. 157–162.

[11] See the articles in the volumes authored and edited by Michel Janssen, John Norton, Jürgen Renn, Tilman Sauer and John Stachel (Janssen *et al.*, 2007a,b); earlier pertinent publications on the subject that we depend on include Norton (1984), Renn and Sauer (1999), Norton (2000); the last article has also been very helpful for the larger thesis of this book.

[12] As in Einstein (1933a), on p. 300 in Einstein (1994).

further advised his audience that if they wanted to learn anything about a scientist's methods, they should "not listen to their words," but "fix [...] attention on their deeds."[13] Einstein himself indicated where to look in his case: his experiences in the discovery of general relativity. We will retrace his path to the field equations, to understand how Einstein came to his methodological position of 1933 and hence, to better understand his later scientific pursuit.

1.1 The dual method and the Zurich notebook

Let us first turn to the debate between Einstein and Max Abraham in 1912 over the latter's proposal for a relativistic theory of gravity. From this debate we will learn more about Einstein's heuristic mind-set at the time when he embarked on his search for the field equations of general relativity.

1.1.1 Einstein contra Abraham: the need "to think physically"

Early in 1912 Abraham published a new dynamical theory of gravity.[14] In order to follow through on the special theory of relativity and its rejection of instantaneous action at a distance, Abraham had adjusted the formulae of Newtonian gravitation theory in the following, somewhat obvious way (for unit mass and acceleration a, with potential ϕ and mass density ρ):

$$\text{Newton:} \qquad\qquad \text{Abraham:}$$

$$-\vec{\nabla}\phi = \vec{a} \quad \text{(I)} \qquad \longrightarrow \qquad -\partial_\mu\phi = a_\mu \quad \text{(I)}$$

$$\vec{\nabla}^2\phi = \kappa\rho \quad \text{(II)} \qquad \longrightarrow \qquad \partial_\mu^2\phi = \kappa\rho. \quad \text{(II)}$$

Because of its apparent mathematical appeal, Einstein was at first quite taken by the theory; he wrote to his intimate friend Michele Besso: "At the first moment (for 14 days!) I [...] was totally 'bluffed' by the beauty and simplicity of his formulas."[15] But he soon saw past the beauty of Abraham's theory.

The first of Abraham's relations (I) implied a variable speed of light. This was not what troubled Einstein; as we saw, he himself contemplated the possibility of a variable speed of light when gravity entered the fray. Abraham, however, initially made free use of the Lorentz transformations in his theory, which according to

---

[13] Ibid., on p. 296 in Einstein (1994).

[14] Abraham (1912a). Our notation deviates slightly from Abraham's; throughout this book, we have tried to stay as close as possible to the notation of primary literature. However, when notation is obsolete compared to today's usage and a change would not alter content we sometimes differ from the notation of sources.

[15] "Im ersten Augenblick (14 Tage lang!) war ich [...] ganz „geblüfft" durch die Schönheit und Einfachheit seiner Formeln." Einstein to Michele Besso, 26 March 1912, Doc. 377 in Klein *et al.* (1993), pp. 435–438, on p. 437; translation as in Norton (2000), p. 142.

Einstein was no longer allowed owing to the variable c.[16] He started pressing this point with Abraham, through private correspondence and also in public. Their debate grew into a true polemic, which took an unpleasant turn when Einstein did not yield when Abraham started questioning the validity of special relativity. Einstein broke away from the public discussion, making clear that he had not changed his mind about the inadequacy of Abraham's theory.[17]

In his private correspondence he indicated how in his opinion Abraham had been led to his crippled theory:

Abraham [...] made some serious mistakes in reasoning so that things are probably incorrect. This is what happens when one operates formally, without thinking physically![18]

Abraham's theory has been created out of thin air, i.e. out of nothing but considerations of mathematical beauty, and is completely untenable. How this intelligent man could let himself be carried away with such superficiality is beyond me.[19]

According to Einstein, Abraham had been misled by mathematical esthetics.

Indeed, in his younger years Einstein had had a lukewarm attitude towards mathematics. For instance, he had been unimpressed by the appearance of Hermann Minkowski's representation of his own special theory, and in 1912 he welcomed the announcement that Paul Ehrenfest would succeed Hendrik Antoon Lorentz in Leiden by congratulating him as "one of the few theorists who has not been robbed of his natural mind by the mathematical epidemic!"[20] This attitude may have softened to a degree when he learnt of impressive new tensor methods the same year, yet, having had a forewarning in Abraham's failure, it is likely that Einstein would have started off on his own search for a relativistic theory of gravity with a cautious attitude towards considerations of mathematical beauty and simplicity.[21]

---

[16] For an infinitesimal transformation: $dx' = (dx - v dt)(1 - v^2/c^2)^{-1/2}$. To integrate this to a full coordinate x', the differential dx' needs to be exact: $\frac{\partial}{\partial t}(1 - v^2/c^2)^{-1/2} = \frac{\partial}{\partial x}[-v(1 - v^2/c^2)^{-1/2}]$. In the case of a static field in which c is a function of the spatial coordinates, this is not fulfilled; see Einstein (1912a), p. 368; see also Norton (2000), pp. 155–157, Renn (2007), p. 315.

[17] See Abraham (1912b), Einstein (1912c), Abraham (1912c), Einstein (1912d). For overviews, see: Norton (2000), pp. 141–144, Klein *et al.* (1995), the editorial note, "Einstein and the static theory," pp. 122–128, and Renn (2007).

[18] "Abraham [hat] bedenkliche Denkfehler [...] gemacht, sodass die Sache wohl unrichtig ist. Das kommt davon, wenn man formal operiert, ohne dabei physikalisch zu denken!" 27 January 1912, Einstein to Heinrich Zangger, Doc. 344 in Klein *et al.* (1993), pp. 394–395, on p. 395; translation as in Norton (2000).

[19] "Abrahams Theorie ist aus dem hohlen Bauche, d.h. aus blossen mathematischen Schönheitserwägungen geschöpft und vollständig unhaltbar. Ich kann gar nicht begreifen, wie sich der intelligente Mann zu solcher Oberflächlichkeit hat hinreissen lassen können." 26 March 1912, Einstein to Michele Besso, Doc. 377 in Klein *et al.* (1993), pp. 435–438, on pp. 436–437; translation as in Norton (2000).

[20] "Sie sind einer der wenigen Theoretiker, denen die Mathematik-Seuche nicht den natürlichen Verstand geraubt hat!" Einstein to Ehrenfest, before 20 June 1912, Doc. 409, pp. 484–486 on p. 484 in Klein *et al.* (1993); see also Frans van Lunteren, "Theoretische fysica als zelfstandig specialisme" (typescript, 2008). On Minkowski, see Pais (1982), p. 152.

[21] Einstein's softening is suggested by a letter to Arnold Sommerfeld: "[...] I have acquired an enormous respect for mathematics" ("[...] ich [habe] grosse Hochachtung für die Mathematik eingeflösst bekommen [...]." 29 October 1912, Doc. 421 in Klein *et al.* (1993), pp. 505–506, on p. 505).

1.1.2 The Zurich notebook: dual method

In 1913 Einstein and Marcel Grossmann jointly published a paper entitled "Entwurf einer verallgemeinerten Relativitätstheorie und einer Theorie der Gravitation" – "Outline of a generalized theory of relativity and of a theory of gravitation."[22] It contained the first tensor-like field equations for gravitation. These were not yet the Einstein equations however. In fact, these field equations were not generally covariant, even though that was initially one of Einstein's demands. We will refer to them as the "Entwurf" ("outline") field equations.

As already mentioned, the genesis of the Entwurf theory has been reconstructed in great detail by studies of notes that Einstein wrote in a notebook in the second half of 1912 and early months of 1913 – the "Zurich notebook."[23] This notebook has revealed that Einstein was mostly working along a two-pronged or dual strategy. It consisted of both a physical and a mathematical approach to finding the appropriate field equations.[24] With this strategy, Einstein was trying to close in on the following question:

What is the appropriate expression $?_{\mu\nu}$ which is formed from the metric and its first and second derivatives and which enters a field equation of the form

$$?_{\mu\nu} = \kappa T_{\mu\nu}, \tag{1.2}$$

with the stress-energy tensor $T_{\mu\nu}$ of matter as the source term on the right-hand side?

The dual method entailed that Einstein abstracted from a set of requirements of a physical nature.

- Newtonian limit: the theory must correspond to Newtonian theory in the case of sufficiently weak and static gravitational fields. Thus, in this limit one must recover a Poisson(-like) equation for one scalar gravitational potential from the field equations.
- Conservation of energy and momentum.

In the bottom-up physical approach Einstein typically tried to find field equations for the metric by generalizing the Poisson equation of Newtonian gravity.[25] Such attempts would take place in a way that guaranteed compliance with the above "correspondence" and "conservation" requirements; the covariance properties of such constructs would however usually be unknown in advance.

[22] The article appeared first as a booklet (Einstein and Grossmann, 1913), and was subsequently published in the *Zeitschrift für Mathematik und Physik*, **62**, pp. 225–259, 1914.

[23] The notebook is thus named as Einstein was working in Zurich at the time. It is transcribed in Klein *et al.* (1995), Doc. 10, pp. 200–269. A facsimile and transcription is also found in Janssen *et al.* (2007a), pp. 313–487; we will follow this last publication's pagination of the notebook. A facsimile version is available online: http://www.alberteinstein.info/ at archival entry no. 3 006.

[24] Also called his "double" strategy; see Janssen *et al.* (2007c), pp. 500–501, Renn and Sauer (1999; 2007).

[25] Another important source for suggestive analogies in the physical approach was the Lorentz–Maxwell theory for electrodynamics. We largely omit this context in the following for brevity; a comprehensive account that includes analogies with Maxwell theory in its discussion of Einstein's dual strategy is Renn and Sauer (2007).

The complementary top-down mathematical approach stemmed from two different requirements.

- The principle of equivalence: observations made in a uniformly accelerated system are equivalent to observations made in a homogeneous gravitational field.
- Generalization of the principle of relativity (i.e. the laws of physics are identical for relatively inertially moving observers) to observers in accelerated motion.

Here, Einstein tried to reason deductively from the more formal, mathematical requirement of general covariance.

- The field equations have to be generally covariant, as an expression of the principle of equivalence and the generalization of the principle of relativity to accelerated observers.

A natural starting point for this approach would be to study a generally covariant object that was known from the mathematics literature.

The dual strategy employed both physical and mathematical approaches: it was an iterative process that began with trying out one approach, and then checking results with the demands or results of the other approach. Einstein would subsequently attempt to accommodate conflicting findings by adjusting his relations, or he would start over again. On occasion, he would try to reassess the relative weight attributed to the above heuristic requirements.

There is of course an element of arbitrariness in the above division of Einstein's heuristics according to academic discipline; the boundaries between fields in the above cases are gradual rather than sharp. Clearly the requirement of general covariance has a physical component – one need only think of Mach's criticism of Newton's bucket and the associated demand of rotation symmetry. Likewise, mathematical background knowledge is involved in the physical demands. Nevertheless, as this chapter will make clear, the above division is helpful in describing and understanding Einstein's overall strategy as he proposed and rejected candidate gravitation tensors. It also throws light on his later assessments of the successes and failures of his attempts.

Finally, today it is not hard to imagine that the above constraints point to the Einstein tensor ($G_{\mu\nu}$) as a candidate for the gravitational term.[26] Both sets of constraints together seem to leave barely an alternative for $?_{\mu\nu}$. Indeed, in his notebook Einstein was initially led to the tensor named after him. However, as we will see, he rejected it as his network of demands and beliefs contained further elements that made it an unattractive option.

---

[26] The Einstein tensor is $G_{\mu\nu} = R_{\mu\nu} - \frac{1}{2}g_{\mu\nu}R$, with $R_{\mu\nu}$ the Ricci tensor and R the Einstein–Hilbert action.

1.1.3 The dual method produces the linearized Einstein tensor

We begin by outlining an attempt from the Zurich notebook in line with Einstein's physical strategy. Generalizing the Poisson equation by replacing Newton's scalar potential with the metric tensor and the Laplacian with contracted four-dimensional differential operators suggested the following object:[27]

$$?_I^{\mu\nu} = g^{\alpha\beta} \frac{\partial^2 g^{\mu\nu}}{\partial x^\alpha \partial x^\beta}. \tag{1.3}$$

In a weak field approximation one easily recovers a d'Alembertian operator acting on the metric; assuming static fields, this becomes again the Laplacian operator.

Einstein considered this object to be a basic component of a candidate for $?_{\mu\nu}$ and thus studied its covariance properties. These were problematic, as $?_I^{\mu\nu}$ seemed not to transform as a tensor. That would rule out the desired generalization of the principle of relativity to accelerated motions. However, one could hope to amend this by adding correction terms to $?_I^{\mu\nu}$, terms quadratic in first-order derivatives of the metric so that the easy recovery of the Laplacian for weak fields would not be disturbed.

Einstein likely wanted to identify appropriate additional terms for $?_I^{\mu\nu}$ by exploring its compliance with energy momentum conservation. He used the following expression for energy momentum conservation for this purpose:[28]

$$\frac{\partial}{\partial x^\mu}(g_{\kappa\nu}T^{\mu\nu}) - \frac{1}{2}\frac{\partial g_{\mu\nu}}{\partial x^\kappa}T^{\mu\nu} = 0. \tag{1.4}$$

Page 13R of the Zurich notebook contains an expression that is similar to (1.4) with $?_I^{\mu\nu}$ instead of $T^{\mu\nu}$:

$$\frac{\partial}{\partial x^\mu}\left(g_{\kappa\nu}g^{\alpha\beta}\frac{\partial^2 g^{\mu\nu}}{\partial x^\alpha \partial x^\beta}\right) - \frac{1}{2}\frac{\partial g_{\mu\nu}}{\partial x^\kappa}g^{\alpha\beta}\frac{\partial^2 g^{\mu\nu}}{\partial x^\alpha \partial x^\beta}. \tag{1.5}$$

Einstein knew that $?_I^{\mu\nu}$ by itself was not a good candidate. He further knew that the gravitational field would produce its own contribution to the energy-momentum content of a system.[29] He would thus have expected that the full field equation would have the form:

$$?_I^{\mu\nu} - \kappa t^{\mu\nu} = \kappa T^{\mu\nu}, \tag{1.6}$$

with $t^{\mu\nu}$ the unknown energy-momentum term of the gravitational field that should be quadratic in first-order derivatives of the metric.

---

[27] See Einstein's Zurich notebook, e.g. pages 07L, 11R, i.e. pp. 390, 392; 407, 409 in Janssen *et al.* (2007a); see also Renn and Sauer (2007), pp. 162–163.

[28] Einstein had set $\sqrt{g} = 1$; the following expression is equivalent to setting the covariant divergence of $T^{\mu\nu}$ to zero.

[29] See Janssen *et al.* (2007c), pp. 550–557.

Presumably Einstein was trying to use expression (1.5) to identify an expression for $t^{\mu\nu}$ and thus uncover the full field equation. He would then have to substitute in the conservation equation (1.4) the gravitational terms $?_I^{\mu\nu} - \kappa t^{\mu\nu}$ for $\kappa T^{\mu\nu}$ and bring terms with $\kappa t^{\mu\nu}$ over to the right-hand side of the conservation equation. This would offer the opportunity to identify an expression for $t^{\mu\nu}$ by comparing terms, that is, identify an expression for $t^{\mu\nu}$ by comparing (1.5) with

$$\frac{\partial}{\partial x^\mu}(g_{\kappa\nu}\kappa t^{\mu\nu}) - \frac{1}{2}\frac{\partial g_{\mu\nu}}{\partial x^\kappa}\kappa t^{\mu\nu}. \tag{1.7}$$

To find the appropriate expression that ought to be identified with $t^{\mu\nu}$, Einstein would have to get rid of the terms in (1.5) that contained third-order derivatives of the metric. Namely, as $t^{\mu\nu}$ was to contain only first derivatives of the metric, expression (1.7) should not have derivatives of higher than the second degree. On page 13R of his notebook, Einstein was indeed engaged in rewriting (1.5) by imposing a certain coordinate condition that would eliminate the undesirable third-order terms, but he did not finish the calculation. Einstein broke off constructing $?^{\mu\nu}$ with $?_I^{\mu\nu}$ at its core when a more promising approach appeared along the mathematical strand.[30]

In any case, we see here Einstein trying his bottom-up, familiar-physics-first approach. When proposing $?_I^{\mu\nu}$ as a candidate for the second derivative term in $?^{\mu\nu}$ he expanded in an apparently straightforward manner on a known relation, the Poisson equation. The expression $?_I^{\mu\nu}$ itself was ruled out as a candidate for $?^{\mu\nu}$, but comparison with energy momentum conservation was expected to produce the terms that, welded together with $?_I^{\mu\nu}$, gave such a candidate.

Marcel Grossmann must at this point have drawn Einstein's attention to the Riemann tensor, $R_{\alpha\mu\nu\beta}$. On the next page of the Zurich notebook, page 14L, Einstein wrote down the Riemann tensor, and right next to it he wrote Grossmann's name (see Figure 1.1). This would have been a natural starting point for the mathematical approach: the Riemann tensor is a generally covariant object constructed from the metric $g_{\mu\nu}$ and its first and second derivatives, and is linear in the second derivatives. But this tensor has four indices, and thus cannot itself be considered a candidate for $?_{\mu\nu}$. On the same page of the Zurich notebook, Einstein constructed the Ricci tensor ($R_{\mu\nu}$), by once contracting the Riemann tensor

$$R_{\mu\nu} = g^{\alpha\beta}R_{\alpha\mu\nu\beta}. \tag{1.8}$$

After having found a generally covariant two index tensor, Einstein had to examine whether it would fit his physical requirements. A difficulty surfaced: the recovery of the Poisson equation in the Newtonian limit appeared to be problematic. The Ricci tensor gives, in the weak field limit, not only the second-order derivative that

---

[30] Janssen *et al.* (2007c), pp. 596–602.

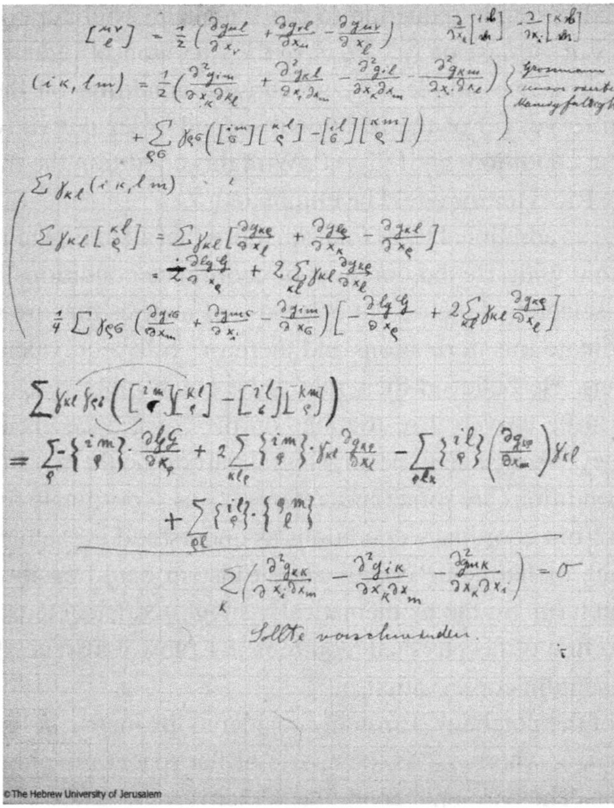

Figure 1.1. Page 14L of the Zurich notebook (by permission of the Albert Einstein Archives, Hebrew University, Jerusalem). Grossmann's name is written next to, what was in Einstein's notation, the Riemann tensor $(i\kappa, lm)$. The Ricci tensor is on the next line, as the contraction of the Riemann tensor with the contravariant metric, written by Einstein as $\gamma_{\kappa l}$; Einstein might have indicated with the question mark following it that he wished to consider the Ricci tensor a candidate for $?_{\mu\nu}$.

$?_I^{\mu\nu}$ also produces, but additional terms. This is due to the fact that the Ricci tensor contains four second-order derivatives:

$$g^{\kappa\lambda}\left(-\frac{\partial^2 g_{\mu\nu}}{\partial x^\kappa \partial x^\lambda} + \frac{\partial^2 g_{\mu\kappa}}{\partial x^\nu \partial x^\lambda} + \frac{\partial^2 g_{\kappa\nu}}{\partial x^\lambda \partial x^\mu} - \frac{\partial^2 g_{\kappa\lambda}}{\partial x^\nu \partial x^\mu}\right). \tag{1.9}$$

Only the first term is of the form $?_I^{\mu\nu}$ of equation (1.3). On the bottom of page 14L, Einstein wrote the other three terms, put them to zero, and explicitly added "should disappear" ("Sollte verschwinden").[31]

---

[31] For the treatment of the Riemann tensor in the Zurich notebook, see Janssen *et al.* (2007c), pp. 603–705, Renn and Sauer (1999), pp. 102–119.

Einstein wanted a simple reduction to the Laplacian operator for a weak and static field, just as in the case of $?_I^{\mu\nu}$. To have the Ricci tensor reduce to a Poisson-like equation in the same simple way, he chose a condition on the coordinates that enabled him to get rid of the additional second-order derivatives: the "harmonic" coordinate condition, $g^{\mu\nu}\Gamma^\alpha_{\mu\nu} = 0$. With this condition the three unwanted derivatives in the Ricci tensor could be eliminated.

This introduced an additional set of four equations that all of Einstein's constraints had to be consistent with. He did not think of coordinate conditions in the modern sense, that is, as selecting a metric out of a class of metrics that are related to one another by coordinate transformations and therefore fully equivalent solutions of the field equations. He believed their role to be much more restrictive: he took such conditions to be universal restrictions on the covariance group of the field equations, and they were taken to be an integral part of the theory. In the historical literature, such conditions are sometimes referred to as "coordinate restrictions," to distinguish them from coordinate conditions as understood in the modern sense.[32]

With the choice of harmonic coordinates, Einstein could easily re-derive the desired Laplacian term. So the mathematically suggested Ricci tensor looked like it would meet the first of his physical requirements. Now it also needed to comply with energy momentum conservation.[33]

On page 19R of the notebook, Einstein considered the metric in linearized form, i.e. $g_{\mu\nu} = \eta_{\mu\nu} + h_{\mu\nu}$, where $\eta_{\mu\nu}$ is the metric of flat spacetime on which $h_{\mu\nu}$ is a small perturbation. The harmonic condition is then:

$$\partial_\mu h^\mu_\beta - \frac{1}{2}\partial_\beta h^\mu_\mu = 0. \tag{1.10}$$

On the same page, Einstein chose the example of dust for the energy-momentum tensor: $T_{\mu\nu} = \rho u_\mu u_\nu$. This gave for the Ricci tensor field equation:

$$\partial^\alpha \partial_\alpha h_{\mu\nu} = \kappa\rho u_\mu u_\nu. \tag{1.11}$$

An easy way to ensure energy momentum conservation with the above field equation is to impose the additional conditions

$$\partial_\mu h^\mu_\nu = 0. \tag{1.12}$$

Together with the harmonic condition (1.10) this produced a problem for Einstein. From the two sets of conditions it followed that the trace of $h_{\mu\nu}$ should be a constant:

$$\partial_\beta h^\mu_\mu = 0. \tag{1.13}$$

---

[32] See for example Janssen *et al.* (2007c), pp. 524–525; Norton (2007) contains a discussion of the Zurich notebook that turns on the distinction between coordinate conditions and coordinate restrictions.

[33] Einstein was not aware of the (contracted) Bianchi identities; see Pais (1982), pp. 274–276, Rowe (2002).

The linearized field equation then implied $T_\mu^\mu = 0$, inconsistent with, for instance, the dust example.

It appears that for this reason Einstein subsequently moved to the (linearized) Einstein tensor as a candidate for $?_{\mu\nu}$. On page 20L of his notebook, he eventually wrote for the linearized field equations an expression that was essentially the same as:[34]

$$\partial^\alpha \partial_\alpha \left(h_{\mu\nu} - \frac{1}{2} \eta_{\mu\nu} h_\beta^\beta \right) = T_{\mu\nu}. \tag{1.14}$$

On the left-hand side, one recognizes the linearized Einstein tensor for weak fields.

Imposing $\partial_\mu T_\nu^\mu = 0$ was now consistent with the harmonic coordinate restriction (1.10). This meant that one no longer had to impose (1.12) and that the problematic constraint on the trace (1.13) had been removed. Yet, the above set of equations again produced an inconsistency: Einstein found that the (linearized) field equations that followed from the mathematically most natural approach did not agree with what he believed to be the right form for the metric of a static mass.

1.1.4 Physics-first prevails and leads to the Entwurf theory

In the case of a field produced by a static mass, Einstein was convinced that all the metric's diagonal spatial components were constant and that the only variable was the g_{00} corner-component:

$$g_{\mu\nu} = \begin{pmatrix} g_{00} & 0 & 0 & 0 \\ 0 & -1 & 0 & 0 \\ 0 & 0 & -1 & 0 \\ 0 & 0 & 0 & -1 \end{pmatrix}. \tag{1.15}$$

We will call this the "Newton metric." We have seen that before Einstein moved to a tensor theory of gravity, in his papers on the static theory from 1911 and 1912 he had considered introducing a variable speed of light. In the Entwurf paper, the step he made to arrive at (1.15) was simply to state that for the case of a field produced by a static mass, instead of a Minkowski metric, where $g_{00} = c^2$ is a constant, one has a g_{00} that varies with the spatial coordinates. Einstein seems to have followed the same argument in the Zurich notebook.[35]

For $?_I^{\mu\nu}$ this static metric gives for weak fields a straightforward reduction to the Poisson equation. However, it was not consistent with the linearized Einstein

---

[34] In his notebook Einstein wrote out three of the components of equation (1.14), i.e. for T_{11}, T_{12} and T_{14}, and indicated by lines and a dot the other components. The constant κ was not written.
[35] Einstein and Grossmann (1913), p. 7; see his Zurich notebook, p. 39L, Janssen *et al.* (2007c), p. 504.

equations. Page 20L of the notebook contained two versions of the (linearized) Einstein equation (1.14), with the second one equivalent to:

$$\partial^{\alpha}\partial_{\alpha}h_{\mu\nu} = T_{\mu\nu} - \frac{1}{2}\eta_{\mu\nu}T^{\beta}_{\beta}.$$ (1.16)

If a non-moving, static mass is curving space then all the $T_{\mu\nu}$ components of the tensor for dust vanish, except for T_{00}. Clearly, (1.15) then cannot be a solution of field equation (1.16), as other diagonal components than g_{00} will also deviate from their Minkowski metric values.

Two pages after writing the linearized Einstein tensor, Einstein reconsidered his static metric. Alarmingly, he found that by giving it up he ran the risk of violating Galileo's law: he, incorrectly, had formulated an expression that entailed that if the static metric had non-trivial diagonal components other than g_{00}, then bodies with different masses would no longer fall at an equal rate in a gravitational field. So Einstein retained the Newton metric (1.15). In the face of such strong physical opposition he rather discarded the linearized Einstein tensor.[36]

With a hindsight bias that should be too comfortable for the historian one sees that to allow fields only of the above form for point masses is too restrictive. For example, the Schwarzschild metric is not of the above form (for future purposes we give its weak field approximation, i.e. for the case that $2m/r \ll 1$):

$$\mathrm{d}s^2 = \left(1 - \frac{2m}{r}\right)\mathrm{d}t^2 - \left(\delta_{ij} + 2m\frac{x_i x_j}{r^3}\right)\mathrm{d}x^i\mathrm{d}x^j.$$ (1.17)

Sticking to (1.15) together with the physical and mathematical requirements implied an over-determination in the search for the field equations, which produced an inconsistency.

Shortly after dismissing the (linearized) Einstein equations, Einstein ended all attempts at finding field equations using the mathematical strategy, and he returned to his physics-first approach. He started once more with the $?^{\mu\nu}_I$-object inspired by the Poisson equation and tried to see again what terms he needed to weld it together with to comply with energy momentum conservation, as he had done before. This attempt produced the set of field equations now referred to as the "Entwurf" field equations.[37]

---

[36] See Janssen *et al.* (2007c), pp. 637–642.
[37] See the Zurich notebook, pp. 26L–26R: Janssen *et al.* (2007a), pp. 466–469, and Janssen *et al.* (2007c), pp. 706–712.

Instead of the Einstein field equations, Einstein and Grossmann published in 1913 a field equation that contained the following object:

$$
?_{\mu\nu}^{\text{Entwurf}} = \frac{1}{\sqrt{-g}} \frac{\partial}{\partial x^\alpha} \left(\sqrt{-g} g^{\alpha\beta} \frac{\partial g_{\mu\nu}}{\partial x^\beta} \right)
$$

$$
- g^{\alpha\beta} g^{\tau\rho} \frac{\partial g_{\mu\tau}}{\partial x^\alpha} \frac{\partial g_{\nu\rho}}{\partial x^\beta} - \frac{1}{2} \frac{\partial g_{\tau\rho}}{\partial x^\mu} \frac{\partial g^{\tau\rho}}{\partial x^\nu} + \frac{1}{4} g_{\mu\nu} g^{\alpha\beta} \frac{\partial g_{\tau\rho}}{\partial x^\alpha} \frac{\partial g^{\tau\rho}}{\partial x^\beta}. \quad (1.18)
$$

This expression, they knew, is not generally covariant.[38] It had however just one term that involved a second derivative of the metric, the same term as $?_{\mu\nu}^{I}$ in (1.3), and indeed, the quadratic terms in (1.18) were included to ensure energy momentum conservation. An obvious advantage of the Entwurf equations was that, when taking the Newtonian limit, one did not have to introduce extra coordinate restrictions to recover the Poisson equation in its, to Einstein, familiar form, and inconsistencies with the Newton metric were therefore not to be expected.

In his notebook, Einstein practically had the equations that he would publish in 1915. He had been led to them by the interplay of the dual method, but they appeared to be inconsistent with his network of physical demands. Einstein discarded the demand for general covariance and in the end opted for that $?_{\mu\nu}$ which lay closest to what was suggested by the Newtonian limit; he let his physical requirements prevail over the mathematically natural. In light of the argument with Abraham, this can hardly seem surprising.

1.2 Conceptual struggles with the Entwurf theory

The lack of general covariance of the Entwurf field equations initially gave Einstein some nagging doubts: if the theory was not generally covariant, to what extent did it generalize the principle of relativity? And, would the theory be at odds with the principle of equivalence? He expressed these doubts in a letter to Hendrik Antoon Lorentz:

I am very glad that you receive our study with such warmth. But regrettably the whole business is still so very tricky, that my confidence in the admissibility of the theory is still rather hesitant. So far, the outline [Entwurf] is only satisfying when it deals with the influence of the gravitational field on other physical processes. In that case, absolute differential calculus allows the formulation of equations that are covariant with respect to arbitrary substitutions. The gravitational field ($g_{\mu\nu}$) appears so to speak as the backbone on

---

[38] Einstein and Grossmann (1913), p. 18.

which everything hinges. *But the gravitational equations themselves unfortunately do not have the property of general covariance.*[39]

However, the conflict between the lack of general covariance in the field equations and the principle of equivalence lost its sting in the course of 1913. The principle of equivalence indeed retained its expression in mathematical general covariance, but Einstein redirected his logic, and tried to reason away the apparent inconsistency. One can roughly characterize his compromise as follows.[40]

- He argued that general covariance, as an expression of the principle of equivalence, should be valid on the level of the line-element from which the equations of motion for material points are derived. Also, there should in fact be generally covariant gravitational field equations of which the Entwurf equations are a specification for a particular set of coordinates.
- But on the basis of the generally covariant field equations one cannot uniquely determine the gravitational potentials $g_{\mu\nu}$ – these field equations are thus not physically interesting. The conservation laws function as a restriction on the coordinates and should reduce the (unknown) generally covariant field equations to the Entwurf field equations. These do determine the gravitational potentials uniquely.[41]

The principle of equivalence would be realized since the expression $\mathrm{d}s^2 = g_{\mu\nu}\mathrm{d}x^\mu\mathrm{d}x^\nu$ is invariant under general coordinate transformations. The equation of motion for an observer that is solely under the influence of gravitational forces, $\delta \int \mathrm{d}s = 0$, is thus also invariant. Consequently, one finds the same equation of motion in the case of either a uniformly accelerated coordinate system or a homogeneous gravitational field. Yet the general covariance of the theory is restricted once the conservation laws are imposed. This was still an improvement over Newtonian absolute space – Einstein wrote to Ernst Mach: "The coordinate system has been tailored to the existing world with the help of the energy law, and looses its nebulous a priori existence."[42]

---

[39] "Ich bin beglückt darüber, dass Sie mit solcher Wärme sich unserer Untersuchung annehmen. Aber leider hat diese Sache doch noch so grosse Haken, dass mein Vertrauen in die Zulässigkeit der Theorie noch ein schwankendes ist. Befriedigend ist der Entwurf bis jetzt, soweit es sich um die Einwirkung des Gravitationsfeldes auf andere physikalische Vorgänge handelt. Denn der absolute Differenzialkalkül erlaubt hier die Aufstellung von Gleichungen, die beliebigen Substitutionen gegenüber kovariant sind. Das Gravitationsfeld ($g_{\mu\nu}$) erscheint sozusagen als das Gerippe, an dem alles hängt. *Aber die Gravitationsgleichungen selbst haben die Eigenschaft der allgemeinen Kovarianz leider nicht.*" 14 August 1913, Einstein to Lorentz, Doc. 467, pp. 546–550, on p. 547 in Klein *et al.* (1993); emphasis as in original. For more on Lorentz's role in the development of general relativity, see Kox (1988).

[40] This position comes clearly to the fore in Einstein (1914a).

[41] The conservation laws followed from taking the divergence of the field equations. They were: $B_\sigma \equiv (\partial^2/\partial x^\alpha \partial x^\nu)(\sqrt{-g}g_{\sigma\mu}g^{\alpha\beta}(\partial/\partial x^\beta)g^{\mu\nu}) = (\partial/\partial x^\nu)(\sqrt{-g}(T_\sigma^\nu + t_\sigma^\nu)) = 0$, where t_σ^ν has been identified in (1.18) as the Entwurf gravitational energy-momentum tensor, and T_σ^ν is the matter energy-momentum tensor. See Einstein and Grossmann (1914), pp. 217–219; see also Norton (1984), p. 133.

[42] "Das Bezugssystem ist der bestehenden Welt mit Hilfe des Energiesatzes sozusagen angemessen und verliert seine nebulose apriorische Existenz." December 1913, Einstein to Ernst Mach, Doc. 495, pp. 583–584, on p. 584 in Klein *et al.* (1993).

The second point above – on the non-uniqueness of the potentials – was under-pinned by the "hole argument." The hole argument showed that with generally covariant field equations one cannot mathematically uniquely determine the potentials $g_{\mu\nu}$, from which Einstein concluded that the physical situation is also not uniquely fixed. The argument went as follows.[43]

Hole argument

One assumes the validity of generally covariant field equations in a spacetime that contains a known matter distribution. In this matter distribution there is a hole H, inside of which there is a vacuum. Inside H one can construct a metric solution g of the field equations, given in the coordinate system x, that is: $g(x)$. One next performs a coordinate transformation that smoothly goes over to the identity on the boundary of H and that transforms the coordinate system x into a new system x'. Solving the field equations in the new coordinate system gives for the metric inside the hole $g'(x')$. In general, $g'(x')$ and $g(x)$ are different mathematical expressions; the functional dependence on the coordinates in each case is in general different. However, one can replace in $g'(x')$ the coordinates x' with coordinates x and in this way construct a solution $g'(x)$. Since $g'(x)$ shares the same functional dependence on the coordinates as $g'(x')$, it is again a solution of the field equations. That means that there are two different solutions $g'(x)$ and $g(x)$ in the same coordinate system x, with the same boundary conditions outside the hole. Einstein contended that these two different solutions represent distinct gravitational fields. Therefore, "happenings in the gravitational field cannot be uniquely determined by generally covariant differential equations for the gravitational field."[44]

If we compare this argument with the ideas from the general theory of relativity, the problem lies in its conclusion, as indeed Einstein came to realize later: $g'(x)$ and $g(x)$ do represent the same physical situation, but its mathematical description is not uniquely determined by general covariant field equations. Einstein later claimed that his original argument was mistaken because it presupposed an independent existence of the coordinate system x, with respect to which one could refer

---

[43] The precise point of Einstein's hole argument has been the issue of some debate; John Stachel (1989b) was the first to point out its current interpretation. The presentation here follows the account in Norton (2007), pp. 751–758, which is primarily based on Einstein's formulation in Einstein (1914c), p. 1067.

[44] "[D]urch allgemein kovariante Differentialgleichungen für das Gravitationsfeld kann das Geschehen in demselben nicht eindeutig festgelegt werden," Einstein (1914c), p. 1067.

to two physically distinct solutions. In 1916 he found that *"real* is physically nothing but the totality of spacetime point coincidences"[45] and these would be the same in the case of either description. In 1914, however, the hole argument convinced Einstein that the mathematical property of general covariance precluded the possibility of finding physically unique solutions, and this made generally covariant field equations from a physical perspective dead wood.

With the hole argument in place, Einstein's confidence in the Entwurf theory got quite a boost. He wrote to his former assistant Ludwig Hopf:

I am now very satisfied with the gravitation theory. The fact that the gravitational equations are not generally covariant, something that quite disturbed me for a while, is unavoidable; it can easily be shown that a theory with generally covariant equations cannot exist if the demand is made that the field is mathematically completely determined by matter.[46]

As to the existence of generally covariant equations, there are some contradictory statements in Einstein's writings. In the late 1913 letter to Hopf above, Einstein denied their existence. Later he expressed the opposite view: in a brief 1914 review of the Entwurf theory, that also included the hole argument, he found the existence of generally covariant field equations principally important, though they were deemed not physically interesting. Einstein's position in that article was indeed that the (unknown) generally covariant equations would be reduced via coordinate conditions into the Entwurf form.[47]

Now could that assertion be true? Are the Entwurf field equations a manifestation of some set of generally covariant field equations, written in a particular coordinate system? The absence of generally covariant field equations to compare the Entwurf equations with gave the theory a contingent character. Max Abraham and another critic, Gustav Mie, pointed out this problematic absence in their publications,[48] but Einstein dismissed such criticism:

So if we, without knowing the covariant equations of the gravitational field, specialize the coordinate system and formulate the gravitational equations exclusively in that special

[45] *"Real* ist physikalisch nichts als die Gesamtheit der raumzeitlichen Punktkoinzidenzen," Einstein to Besso, 3 January 1916, Doc. 178, pp. 234–236, on p. 235 in Schulmann *et al.* (1998); emphasis in original. John Earman and John Norton have used Einstein's post-1915 position on the hole argument to argue against spacetime *substantivalism*, i.e. the position that ascribes the spacetime manifold a physical reality in itself (Earman and Norton, 1987). Their argument essentially repeats the hole construction for active transformations on spacetime points; in generally covariant theories, such a transformation should lead to observationally equivalent situations and leave the laws of the theory invariant. This then poses a problem for the substantivalist, who would maintain that the alteration of the spacetime points does constitute a change in the physical system. On the hole argument, see also Rynasiewicz (1994).

[46] "Mit der Gravitationstheorie bin ich nun sehr zufrieden. Die Thatsache, dass die Gravitationsgleichungen nicht allgemein kovariant sind, welche mich vor einiger Zeit noch so ungemein störte, hat sich als unumgänglich herausgestellt; es lässt sich einfach beweisen, dass eine Theorie mit allgemein kovarianten Gleichungen nicht existieren kann, falls verlangt wird, dass das Feld durch die Materie mathematisch vollständig bestimmt wird." 2 November 1913, Einstein to Hopf, Doc. 480, pp. 562–563, on p. 562 in Klein *et al.* (1993).

[47] See Einstein (1914a).

[48] Abraham (1914), p. 18, Mie (1914a) and in particular Mie (1914b), on p. 176, reprinted in Renn and Schemmel (2007b), pp. 699–728, on p. 727; for an introduction to Mie's approach to gravity, see Smeenk and Martin (2007).

coordinate system, then we expose the theory to [...] the objection that the formed equations are perhaps without any physical content. But nobody can seriously believe this objection to be justified in the case at hand.[49]

Deriving the Entwurf equations from generally covariant equations would amount to a deductive check of the Entwurf theory. But despite the lack of such a verification Einstein already had enough reasons to believe in the Entwurf equations: his arguments abstracted from his network of physical demands.

The Newton metric (1.15) is of course a solution of the Entwurf field equations, and Einstein used this metric to calculate for the first time the perihelion shift of Mercury. The answer was a factor of 2.4 off from the observed value, and he did not publish the result.[50] Surprisingly, Einstein did not seem too much taken aback by this incorrect result. He found the faulty value in June 1913, and there are ample occasions after that month in which he expressed confidence in the theory. Two years later, when Einstein finally did abandon the Entwurf theory, he indeed mentioned the discrepancy as one of his motivations.[51]

Early in 1914 Einstein claimed that he could prove that the Entwurf field equations held in every coordinate system that complied with the conservation law. This was found only after great exertions, as one learns from a letter that Einstein wrote to Besso; the same letter further suggests that Einstein himself thought of his searches as following two routes:

At the moment I do not feel like working, for I had to struggle horribly to discover what I described above. The general theory of invariants was only an impediment. The direct route proved to be the only feasible one. It is only difficult to understand why I had to grope around for so long before I found what was so near at hand.[52]

In the search for the Entwurf field equations, too, the mathematical side of his dual strategy had been an obstacle that inhibited progress and was eventually cast aside, in favor of a "physics first" approach.

Yet, the fact that Einstein had let the physical requirements prevail in his initial move to the Entwurf theory did not mean that he had distanced himself from the

---

[49] "Wenn wir also, ohne die allgemein kovarianten Gleichungen des Gravitationsfeldes zu kennen, das Bezugssystem spezialisieren und die Feldgleichungen der Gravitation nur für die speziellen Bezugssysteme aufstellen, so setzen wir die Theorie keinem anderen Einwande aus als demjenigen, dass die aufgestellten Gleichungen vielleicht ohne jeden physikalischen Inhalt sein könnten. An eine Berechtigung dieses Einwandes wird aber im vorliegenden Falle niemand im Ernst denken." As in Einstein (1914a), p. 178.

[50] The calculation was done together with Michele Besso. The Newton metric (1.15) with $g_{00} \sim 1/r$ is written in equation 4 in the so-called "Einstein-Besso" manuscript: Document 14 in Klein *et al.* (1995), on p. 360. The Einstein equations do reproduce the observed value, so one can conclude that the Entwurf equations cannot be the Einstein equations for a particular set of coordinates.

[51] See for example the letter to Sommerfeld, 28 November 1915, Doc. 153, pp. 206–209 in Schulmann *et al.* (1998); for an explanation of Einstein's little worry over the Mercury result, see Renn and Sauer (2007), pp. 261–262.

[52] "Besonders arbeitslustig bin ich jetzt nicht, denn ich habe mich schauerlich plagen müssen, um die obige Sache zu finden. Die allgemeine Invariantentheorie wirkte nur als Hemmnis. Der direkte Weg erwies sich als der einzig gangbare. Unbegreiflich ist nur, dass ich solange tasten musste, bevor ich das Nächstliegende fand." 10 March 1914, Einstein to M. Besso, Doc. 514, pp. 603–604 on p. 604 in Klein *et al.* (1993).

dual method, nor that he had become disinterested in arguments based on mathematical deduction.[53] The dual method was still Einstein's ideal. In general terms it is reflected in his 1914 inaugural address in Berlin:

The methodology of the theoretician mandates implicitly that he use as his basis general assumptions, so-called principles, from which he can then deduce conclusions. His activity, therefore, has two parts: first, he has to ferret out these principles, and second, he has to develop the conclusions that can be deduced from these principles. His school provides him with excellent tools with which to fulfill the second-named task. [...] But the former task, namely to establish these principles which can serve as the basis of his deductions, is one of a completely different kind. Here there is no learnable, systematically applicable method which would lead him to the objective. The researcher must rather eavesdrop on nature to become privy to these general principles, by recognizing in larger sets of experiential facts certain general traits that can then be strictly and precisely formulated. [...]

We have determined that inductive physics has questions for deductive physics and vice versa; and eliciting the answers will require the application of our utmost efforts. May we, by means of united efforts, soon succeed in advancing toward conclusive progress![54]

In the same lecture, Einstein asked for consideration – perhaps out of modesty, or to dampen too high-strung expectations – should he appear with "meager" results before his new fellow Academy members (to be sure, he had been wooed to Berlin by Max Planck and Walther Nernst).[55] He admitted that Planck's quantum still lacked any formal understanding, and that the Entwurf theory had not been experimentally validated. These examples had thus so far eluded a solid capture through Einstein's ideal methodology: the interplay between physics-oriented induction from experience on the one hand, and mathematical deduction on the other – an

---

53 In Einstein's 1914 assessment of a gravity theory proposed by Gunnar Nordström, a rival to his Entwurf theory, he expressed appreciation for this kind of argument. Together with Adriaan Fokker, he wrote: "[O]ne can arrive at Nordström's theory through purely formal considerations, i.e. without assistance of further physical hypotheses, by starting from the principle of the constancy of the velocity of light. Therefore it seems to us that this theory should be preferred over all other gravitation theories that retain this principle." ("[M]an [kann] bei Zugrundelegung des Prinzips von der Konstanz der Lichtgeschwindigkeit durch rein formale Erwägungen, d.h. ohne Zuhilfenahme weiterer physikalischen Hypothesen zur Nordströmschen Theorie gelangen. Es scheint uns deshalb, dass diese Theorie allen anderen Gravitationstheorien gegenüber, die an diesem Prinzip festhalten, den Vorzug verdient.") As in Einstein and Fokker (1914), p. 328. According to Einstein, the Entwurf theory had a variable speed of light, and was still to be preferred over Nordström's work. For more on this work and its reception, see Norton (1992).

54 "Die Methode des Theoretikers bringt es mit sich, dass er als Fundament allgemeine Voraussetzungen, sogenannte Prinzipe, benutzt, aus denen er Folgerungen deduzieren kann. Seine Tätigkeit zerfällt also in zwei Teile. Er hat erstens jene Prinzipe aufzusuchen, zweitens die aus den Prinzipen fliessenden Folgerungen zu entwickeln. Für die Erfüllung der zweiten der genannten Aufgaben erhält er auf der Schule ein treffliches Rüstzeug. [...] Die erste der genannten Aufgaben, nämlich jene, die Prinzipe aufzustellen, welche der Deduktion als Basis dienen sollen, ist von ganz anderer Art. Hier gibt es keine erlernbare, systematisch anwendbare Methode, die zum Ziele führt. Der Forscher muss vielmehr der Natur jene allgemeinen Prinzipe gleichsam ablauschen, indem er an grösseren Komplexen von Erfahrungstatsachen gewisse allgemeine Züge erschaut, die sich scharf formulieren lassen. [...] Wir haben festgestellt, dass die induktive Physik an die deduktive und die deduktive an die induktive Fragen stellt, deren Beantwortung die Anspannung aller Kräfte erfordert. Möge es bald gelingen, durch vereinte Arbeit zu endgültigen Fortschritten vorzudringen!" As in Einstein (1914b), pp. 740–742.

55 "[...] wenn Ihnen die Früchte meiner Bemühungen als ärmliche erscheinen werden," (Einstein, 1914b, p. 739); on the Berlin offer, see for example Stern (1999), pp. 111–112, Renn (2006), pp. 69–77.

interplay by which one should close in on the correct theory, resembling the back and forth between physical and mathematical approaches in the Zurich notebook. Even though the Entwurf field equations had turned out to be problematic from an observational point of view, physical arguments had however superseded "deductive" reasoning on that occasion.

Einstein still intended to place the Entwurf theory on firmer mathematical ground. At one moment he claimed to have demonstrated that its Lagrangian is uniquely fixed by the requirement that it be invariant under the remaining symmetries of the field equations, combined with the demand of energy momentum conservation. After formulating the result, he believed that he had now arrived "in a purely formal way, i.e. without directly drawing on our physical knowledge of gravitation, at completely definite field equations."[56] His arguments were flawed, however, as was soon pointed out to him by the Italian mathematician Tullio Levi-Civita.[57]

Einstein did not remain committed to the compromise of the Entwurf theory between the principle of equivalence and lack of general covariance that we have outlined earlier. At a certain point he hoped that among the remaining symmetries of the Entwurf field equations were also non-linear transformations of the coordinates, among them transformations to a rotating coordinate system. In that case, for the sake of consistency, any remaining symmetries in the Entwurf theory should be present in both the field equations and the coordinate restrictions (or conservation laws). The second and last paper that Einstein and Grossmann wrote together claimed that there were indeed a number of non-linear transformations which satisfy this requirement.[58] They showed that in principle symmetry transformations of the Entwurf field equations are also symmetries of the coordinate restrictions. However, they did not show whether these symmetries included non-linear transformations. Einstein nevertheless repeated this claim in a letter to Besso, and explicitly mentioned transformations to rotating systems.[59]

Einstein checked at least twice whether Minkowski space would still be a solution of the Entwurf equations after a transformation to a rotating system. An early calculation was done in June 1913 and partly due to some trivial but crucial

---

[56] "[...] auf rein formalem Wege, d.h. ohne direkte Heranziehung unserer physikalischen Kenntnisse von der Gravitation, zu ganz bestimmten Feldgleichungen gelangt," Einstein (1914c), p. 1076. The attempt was introduced by words that again reflect a dual approach: "[A] kaleidoscopic mixture of postulates from physics and mathematics has been introduced and used as heuristical tools [in the generalization of the theory of relativity]; as a consequence it is not easy to see through and characterize the theory from a formal mathematical point of view [...]. The primary objective of this paper is to close this gap." ("Als heuristische Hilfsmittel sind bei jenen Untersuchungen in bunter Mischung physikalische und mathematische Forderungen verwendet, so dass es nicht leicht ist, [...] die Theorie vom formal mathematischen Standpunkte aus zu übersehen und zu charakterisieren. Diese Lücke habe ich durch die vorliegende Arbeit in erster Linie ausfüllen wollen.") As in Einstein (1914c), p. 1030; see also Renn and Sauer (2007), pp. 257–260.

[57] For more on their discussions, see Cattani and de Maria (1989).

[58] See Einstein and Grossmann (1914), in particular p. 216.

[59] See Norton (1984), pp. 132–133.

errors, Einstein incorrectly answered the question in the affirmative.[60] When he returned to the problem in September 1915, however, he drew the conclusion that the Minkowski metric in rotating coordinates was not a solution of the Entwurf field equations. That would imply that the theory was not invariant under rotations. The realization now upset him greatly: Einstein reported to the astronomer Erwin Freundlich that he had come across a "logical contradiction" that "enormously galvanized" him.[61]

Much distressed, Einstein would soon give up on the troubled theory. His reasons were threefold, as he wrote to Arnold Sommerfeld:

- I proved that the gravitational field on a uniformly rotating system does not satisfy the gravitational field equations.
- The motion of the perihelion of Mercury came out as $18''$ instead of $45''$ per century.
- The covariance argument in my paper of last year does not give the Hamiltonian function H. When suitably generalized it allows an arbitrary H.[62]

1.3 November 1915: mathematics produces the covariant Einstein equations

Einstein quickly got hold of the way out of the Entwurf quagmire: he returned to the mathematical strategy and the requirement of general covariance that he had abandoned in the Zurich notebook. The four publications from November 1915, submitted on the 4th, 11th, 18th and 25th, document the rapid capture of the final theory, effected by his enhanced understanding of both the mathematics and the physics of the problem developed in the intervening years. Now, in just six weeks, Einstein cast aside the Entwurf theory, went back to the covariant Riemann tensor and deduced the right value for the perihelion shift of Mercury. As John Norton has put it: "Einstein's reversal was his Moses that parted the waters and led him from bondage into the promised land of general relativity."[63]

In the first November paper, Einstein wrote:

[...] I completely lost trust in the field equations I had chosen and looked for a way to restrict the possibilities in a natural manner. Thus I went back to the requirement of a more

---

[60] This calculation is included in the Einstein–Besso manuscript, Doc. 14 in Klein *et al.* (1995).

[61] "[...] logischen Widerspruch [...], [der] mich ungeheuer elektrisiert." Einstein to Freundlich, 30 September 1915, Doc. 123, pp. 177–178, on p. 177 in Schulmann *et al.* (1998). For more on the problem of rotation in the Entwurf theory, see Janssen (1999); an updated account is found in Janssen (2007).

[62] "1. Ich bewies, dass das Gravitationsfeld auf einem gleichförmig rotierenden System den Feldgleichungen nicht genügt. 2. Die Bewegung des Merkur-Perihels ergab sich zu $18''$ statt $45''$ pro Jahrhundert. [Observations at that time had actually given a value of $41'' \pm 2''$ per century (Earman and Janssen, 1993).] 3. Die Kovarianzbetrachtung in meiner Arbeit vom letzten Jahre liefert die Hamilton-Funktion H nicht. Sie lässt, wenn sie sachgemäss verallgemeinert wird, ein beliebiges H zu." Einstein to A. Sommerfeld, 28 November 1915, Doc. 153, pp. 206–209, on pp. 206–207 in Schulmann *et al.* (1998), translation as in Stachel (1989b).

[63] Norton (2000), p. 152. The four November publications are: Einstein (1915a) 4 November, Einstein (1915b) 11 November, Einstein (1915c) 18 November and Einstein (1915d) 25 November.

general covariance of the field equations, which I had left only with a heavy heart when I worked together with my friend Grossmann. In fact we had then already come quite close to the solution of the problem given in the following.[64]

This article did not yet contain the Einstein equations, but it came close. His reasoning started again with the Riemann and Ricci tensors. He divided the latter, $R_{\mu\nu}$, into two terms:

$$R_{\mu\nu} = S_{\mu\nu} + \tilde{S}_{\mu\nu}. \tag{1.19}$$

The left-hand side of the gravitational field equations (i.e. $?_{\mu\nu}$) was to be given by $S_{\mu\nu}$:

$$S_{\mu\nu} \equiv \partial_\alpha \Gamma^\alpha_{\mu\nu} - \Gamma^\beta_{\mu\alpha}\Gamma^\alpha_{\beta\nu} = \kappa T_{\mu\nu}. \tag{1.20}$$

These equations were not generally covariant, but they were covariant for transformations that left $\sqrt{-g}$ invariant.

Possibly, Einstein proposed (1.20) instead of a field equation with the Ricci tensor because of the straightforward reduction to the Poisson equation that it entailed in the case of the Newton metric (1.15).[65] For static and weak fields (1.20) produces a Poisson-like equation under the condition:

$$\partial_\alpha g^{\alpha\mu} = 0, \tag{1.21}$$

which is consistent with (1.15). Einstein would then have had five coordinate restrictions on his hands: (1.21) and the demand that only transformations that leave $\sqrt{-g}$ invariant are allowed. Both sets of conditions were consistent, which he could show by using the conservation laws.[66]

Consistency between the field equations and the invariance requirement for $\sqrt{-g}$ entailed a further condition:

$$\partial_\alpha \left(g^{\alpha\beta} \frac{\partial}{\partial x^\beta} \ln \sqrt{-g} \right) = \kappa T^\mu_\mu, \tag{1.22}$$

from which on 4 November Einstein drew the conclusion that $\sqrt{-g}$ cannot be a constant, since otherwise one would have a severe constraint on the matter sources:

---

[64] "[Ich verlor] das Vertrauen zu den von mir aufgestellten Feldgleichungen vollständig und suchte nach einem Wege, der die Möglichkeiten in einer natürlichen Weise einschränkte. So gelangte ich zu der Forderung einer allgemeineren Kovarianz der Feldgleichungen zurück, von der ich vor drei Jahren, als ich zusammen mit meinem Freunde Grossmann arbeitete, nur mit schwerem Herzen abgegangen war. In der Tat waren wir damals der im nachfolgenden gegebenen Lösung des Problems bereits ganz nahe gekommen" (Einstein, 1915a, p. 778), translation as in Norton (2000), p. 150.

[65] This suggestion originates with John Norton (1984), p. 143, see also Renn and Sauer (2007), p. 272; in a different publication with Michel Janssen, Renn is however critical of Norton's scenario: see Janssen and Renn (2007), pp. 843, 879.

[66] On the Newtonian limit of the 4 November theory and its consistency with unimodular transformations, see Einstein (1915a), p. 786.

$T_\mu^\mu = 0$. Such a constraint implied that all matter should consist of electrody-namic fields. This observation need not necessarily have dismissed the constancy of $\sqrt{-g}$, as turn-of-the-century physics had been under the influence of the so-called electromagnetic view of nature. This view, promoted by many, including for instance Mie, held that all matter was indeed of electromagnetic origin.[67] The electromagnetic world view had been on the decline, however, since the advent of the quantum hypothesis, and it is therefore not surprising that Einstein chose to avoid the implications of a constant $\sqrt{-g}$.

Nevertheless, on 11 November, only a week later, Einstein reversed his position and subscribed to the above conclusion regarding the exclusively electromagnetic constitution of matter.[68] He had noticed that under the coordinate condition

$$\sqrt{-g} = 1, \tag{1.23}$$

the $\tilde{S}_{\mu\nu}$ from (1.19) vanished. This meant that with (1.23) he could write $?_{\mu\nu} = R_{\mu\nu}$ and still retain all the machinery of the 4 November theory. The gain was that the actual field equations were now finally generally covariant. That this achievement implied that all matter needed to be of electromagnetic origin – so that nature through (1.22) produced the correct coordinates – was but a small price to pay, or in Einstein's words, "an admittedly bold additional hypothesis."[69]

It is evident which of the two strands of the dual method was now leading the way: mathematical deduction. Einstein did not have a problem proclaiming all matter to be electromagnetic in origin, since this would finally deliver general covariance and give the entire theory "an even more logically strict structure."[70] On 4 November he had already praised the magic of mathematics:

The magic of this theory can hardly escape anyone who has really understood it: it signifies a true triumph for the method of general differential calculus of Gauss, Riemann, Christoffel, Ricci and Levi-Civita.[71]

Returning to the mathematically natural Riemann tensor would not disappoint him. In the week after 11 November, Einstein redid the calculation of Mercury's perihelion shift. Subsequently, he must have realized that (1.23) cannot be consistent with the Newton metric (1.15), but this time he did not falter: he finally put aside (1.15)

---

[67] On the electromagnetic world view, see McCormmach (1970), Seth (2004).

[68] Einstein included the electromagnetic world view in his motivation for this conclusion, see Einstein (1915b), pp. 799–800.

[69] "[…] einer allerdings kühnen zusätzlichen Hypothese über die Struktur der Materie" (Einstein, 1915b, p. 799). The trace of the stress tensor of the gravitational field need not vanish, according to Einstein, so one could effectively still substitute source terms like $\rho u^\mu u^\nu$, but should account for such terms as due to gravitational stress-energy.

[70] "[…] ein noch strafferer logischer Aufbau," (Einstein, 1915b, p. 799).

[71] "Dem Zauber dieser Theorie wird sich kaum jemand entziehen können, der sie wirklich erfasst hat; sie bedeutet einen wahren Triumph der durch Gauss, Riemann, Christoffel, Ricci und Levi-Civiter [sic] begründeten Methode des allgemeinen Differentialkalküls," (Einstein 1915a, p. 779).

and allowed the spatial components g_{ij} to be functions of the coordinates too. The field equations that Einstein used in the 18 November paper to calculate Mercury's motion gave him a linearized version of the metric later known as the Schwarzschild metric, i.e. (1.17), instead of the Newton metric.[72] He arrived at the correct value for the perihelion shift: Einstein found 43″, which confirmed his desired 45″ ± 5″ per century (since then the measured values have closed in on 43″).[73]

Einstein's calculation was of crucial significance in the final development. He was no longer constrained by the Newton metric and its reduction to the Poisson equation for weak fields. This meant that he could also contemplate different generally covariant field equations than the equations he had published on 11 November. Einstein realized that by adding a trace term T_μ^μ to the source terms he could remove the constraint on matter (1.22) altogether and still retain consistency with the conservation laws. With (1.22) out of the way, matter also no longer needed to consist exclusively of electromagnetic fields. On 25 November Einstein thus finally wrote:

$$R_{\mu\nu} = \kappa \left(T_{\mu\nu} - \frac{1}{2} g_{\mu\nu} T \right) \tag{1.24}$$

which is equivalent to

$$R_{\mu\nu} - \frac{1}{2} g_{\mu\nu} R = \kappa T_{\mu\nu}. \tag{1.25}$$

At last the final field equations had been found. "My wildest dreams have been fulfilled. *General* covariance. Perihelion motion of Mercury wonderfully exact." – "I was beside myself for several days in joyous excitement."[74]

It had been the reversion to the mathematical requirement of general covariance that finally triggered salvation and that in the end relieved Einstein of his prejudices regarding the Newtonian limit. This was most surprising to Einstein. He expressed his amazement at this new development to Besso:

Read the articles! They bring the final salvation from misery. The most joyful aspect is the accordance between the perihelion motion and general covariance, the most striking however the circumstance that Newton's theory for the *field* is already incorrect at the first order of the equation (the terms g_{11}–g_{33} arise). The simplicity of Newton's theory is only

---

[72] In fact, Karl Schwarzschild (1916) used the November 11 field equations to find the exact solution for the field of a point mass named after him. In vacuum these field equations are equivalent to the final Einstein field equations. For more on the early history of the Schwarzschild solution, see Eisenstaedt (1982, 1989a).

[73] Einstein (1915c). For the history of the Mercury problem, see Earman and Janssen (1993).

[74] "Die kühnsten Träume sind nun in Erfüllung gegangen. *Allgemeine* Kovarianz. Perihelbewegung des Merkur wunderbar genau." 10 December 1915, Einstein to Besso, Doc. 162, p. 218 in Schulmann *et al.* (1998), emphasis in the original. "Ich war einige Tage fassungslos vor freudiger Erregung." 17 January 1916, Einstein to Ehrenfest, Doc. 182, pp. 242–244, on p. 244 in Schulmann *et al.* (1998), translations as in Norton (2000).

due to the fact that the $g_{11}-g_{33}$ do not arise in the first approximation to the equations of motion for a point mass.[75]

During the Mercury calculation he had noticed that his prejudice for the static metric (1.15) was already incorrect in the lowest order. However, in the same order the geodesic equation for slow moving point masses contains just one gravitational potential, also if the metric is of the form (1.17).[76]

Einstein realized that his earlier generalization on the basis of Newton's theory that had led him to (1.15) had in fact led him astray, and that the path indicated by the mathematics of general covariance had been the correct one all along. He wrote to Sommerfeld:

You should not be mad at me that only today I reply to your friendly and interesting letter. The last month has been the most exciting and demanding time of my life; in any case also the most successful. I could not think of writing. [...]

After all faith in the results and methods of the earlier theory had dwindled, I clearly saw that only by taking up the general theory of covariants, that is, by taking up Riemann's covariant, a satisfying solution could be found. [...] Of course it is easy to write down these generally covariant equations [i.e. (1.24)]. But it is hard to see that they are the generalization of Poisson's equations and not easy to see that they allow satisfaction of the conservation laws. [...]

With Grossmann I had considered these equations three years earlier, up to the second term on the right hand side, but at the time I came to the incorrect conclusion that they did not give the Newtonian approximation. The key to this solution was the realization that not

$$g^{l\alpha}\frac{\partial}{\partial x^m}g_{\alpha i} \qquad (1.26)$$

but the related Christoffel symbol $\left\{{im \atop l}\right\}$ should be taken as the natural expression for the "components" of the gravitational field. Once one has come to this conclusion, then the above equation [i.e. (1.24)] is the simplest conceivable, since one is not tempted to convert it by multiplying out the symbols for the sake of general interpretation.

---

[75] "Lies die Abhandlungen! Sie bringen die endgültige Erlösung aus der Misere. Das Erfreulichste ist das Stimmen der Perihelbewegung und die allgemeine Kovarianz, das Merkwürdigste aber der Umstand, dass Newtons Theorie *des Feldes* schon in Gl. 1. Ordnung unrichtig ist (auftreten der $g_{11}-g_{33}$). Nur der Umstand, dass die $g_{11}-g_{33}$ nicht in den ersten Näherung der Bew. Gl. des Punktes auftreten, bedingt die Einfachheit von Newtons Theorie." 21 December 1915, Einstein to M. Besso, Doc. 168, p. 223 in Schulmann *et al.* (1998), emphasis in the original. Einstein had sent reprints of his papers earlier on the tenth of December to Besso, see Doc. 162 in Schulmann *et al.* (1998).

[76] See Renn and Sauer (2007), pp. 280–285, Earman and Janssen (1993), pp. 145–146.

I experienced a glorious moment when not only Newton's theory in the first order approximation, but also the perihelion motion of Mercury (43″ per century) in the second order approximation were retrieved.[77]

Upon Einstein's return to "Riemann's covariant," a redefinition of the gravitational field had been necessary to deliver him from the prejudices of the Entwurf theory, according to the letter to Sommerfeld. In that theory, (1.26) had played the role of the gravitational field in the field equations. Just as in electrodynamics the Langrangian is given by $-\frac{1}{4}F^{\mu\nu}F_{\mu\nu}$, the Entwurf Langrangian was equivalent to:[78]

$$H = -\frac{1}{2}g^{\mu\nu}\left(g^{\alpha\sigma}\frac{\partial g_{\sigma\beta}}{\partial x^{\mu}}\right)\left(g^{\beta\tau}\frac{\partial g_{\tau\alpha}}{\partial x^{\nu}}\right). \qquad (1.27)$$

Einstein had developed a variational formalism for the Entwurf theory in the course of 1914 when trying to understand the relation between the theory's covariance properties and the conservation principle. Along with his return to the Riemann tensor, he likely further found a motivation for the first field equations of November 1915 (1.20) by replacing in the above Lagrangian (1.27) the Entwurf definition for the field (1.26) with the Christoffel symbols. Variational calculus would then have led him straight to equation (1.20).[79] Either method gave equations that were the "simplest conceivable," and not some field equation that had been contrived to suit a particular limit.

---

[77] "Sie dürfen mir nicht böse sein, dass ich erst heute auf Ihren freundlichen und interessanten Brief antworte. Aber ich hatte im letzten Monat eine der aufregendsten, anstrengendsten Zeiten meines Lebens, allerdings auch der erfolgreichsten. Ans Schreiben konnte ich nicht denken. [...] Nachdem so jedes Vertrauen im Resultate und Methode der früheren Theorie gewichen war, sah ich klar, dass nur durch einen Anschluss an die allgemeine Kovariantentheorie, d.h. an Riemanns Kovariante, eine befriedigende Lösung gefunden werden konnte. [...] Es ist natürlich leicht, diese allgemein kovarianten Gleichungen hinzusetzen, schwer aber, einzusehen, dass sie Verallgemeinerungen von Poissons Gleichungen sind, und nicht leicht, einzusehen, dass sie den Erhaltungssätzen Genüge leisten. [...] Diese Gleichungen [i.e. (1.24)] hatte ich schon vor 3 Jahren mit Grossmann erwogen, bis auf das zweite Glied der rechten Seite, war aber damals zu dem Ergebnis gelangt, dass sie nicht Newtons Näherung lieferte, was irrtümlich war. Den Schlüssel zu dieser Lösung lieferte mir die Erkenntnis, dass nicht $g^{l\alpha}(\partial/\partial x^{m})g_{\alpha i}$ sondern die damit verwandten Christoffel'schen Symbole $\{{}^{im}_{l}\}$ als natürlichen Ausdruck für die „Komponente" des Gravitationsfeldes anzusehen ist. Hat man dies gesehen, so ist die obige Gleichung [i.e. (1.24)] denkbar einfach, weil man nicht in Versuchung kommt, sie behufs allgemeiner Interpretation umzuformen durch Ausrechnen der Symbole. Das Herrliche, was ich erlebte, war nun, dass sich nicht nur Newtons Theorie als erste Näherung, sondern auch die Perihelbewegung des Merkur (43″ pro Jahrhundert) als zweite Näherung ergab." 28 November 1915, Einstein to Sommerfeld, Doc. 153, pp. 206–209 in Schulmann *et al.* (1998).

[78] Janssen and Renn (2007), p. 869.

[79] This scenario has been developed by Michel Janssen and Jürgen Renn (2007). They in fact deny, contrary to Einstein's own pronouncements, and e.g. the account in Norton (2000), that Einstein had returned to the mathematical strategy centered on general covariance when first finding equations (1.20). Janssen and Renn rather point to the above analogy with electrodynamics and change in the field's definition, augmented with the variational formalism, as the relevant context of discovery, and identify this as exclusively a success of Einstein's physical strategy. In a different publication by Renn, together with Tilman Sauer (2007, see their pp. 270–274), Einstein's 1915 turn to (1.20) is however qualified as a return to the mathematical strategy; the above scenario is included in that account as well. Janssen (2005) nevertheless insists that Einstein found the field equations "at the end of a bumpy road through physics" (p. 82).

In the beginning of the letter above, Einstein apologized to Sommerfeld for not having written sooner. Einstein had virtually suspended his correspondence in the frantic November month, except for his exchanges with mathematician David Hilbert. After having heard Einstein lecture on the Entwurf theory in Göttingen, Hilbert had also found the trail to the relativistic gravity theory – which may have spurred Einstein on in the rapid succession of papers submitted to the Berlin Academy. Until some years ago popular belief had it that Hilbert had included the final field equations in a paper submitted on 20 November, just five days before Einstein submitted his last November article. Archival research has however shown that he did not fully anticipate Einstein:[80] Hilbert had been close, but an explicit expression of the final field equations is absent from a recovered first set of proofs of his paper. Hilbert altered the text of these proofs after Einstein's work came out. The final version of Hilbert's paper appeared as late as 31 March 1916, but the dateline of the article still said no more than that it had been "submitted in the meeting of the 20th of November,"[81] suggesting that Hilbert had beaten Einstein at the finishing line (even if the article did cite Einstein's final 25 November publication on its first page). This course of events initially led to some friction between Einstein and Hilbert, but they soon reconciled and Hilbert always gave Einstein full credit for his work.[82]

1.4 Conclusion: general relativity and Einstein's methodological lesson

The letters that Einstein wrote in the winter months of 1915 and 1916, like the one to Sommerfeld quoted earlier, document his immediate emotions and perspective on what had been the decisive steps. These experiences, after having sunk in, would play a major role in Einstein's later methodological and scientific outlook. For example, in his 1949 autobiography he wrote:

I have learned [...] from the theory of gravitation: no collection of empirical facts however comprehensive can ever lead to the formulation of such complicated equations. A theory can be tested by experience, but there is no way from experience to the construction of a theory. Equations of such complexity as the equations of the gravitational field can be found only through the discovery of a logically simple mathematical condition that determines the equations completely or almost completely. Once one has obtained those sufficiently strong formal conditions, one requires only little knowledge of facts for setting up a theory; in the case of the equations of gravitation it is the four-dimensionality and the symmetric tensor

---

[80] See Corry *et al.* (1997).

[81] "Vorgelegt in der Sitzung vom 20. November 1915" (Hilbert, 1915), p. 895. Offprints of Hilbert's work were already circulating by mid-February.

[82] The published article (Hilbert, 1915) and its proofs are available online: http://echo.mpiwg-berlin. mpg.de/content/modernphysics/hilbert. For context and a nuanced perspective on the Einstein–Hilbert priority issue, see Sauer (1999, 2005).

as expression for the structure of space which, together with the invariance concerning the continuous transformation-group, determine the equations almost completely.[83]

Almost completely indeed: textbooks on relativity theory use three more conditions to fix the Einstein equations completely.[84] First, dimensional grounds suggest that the Einstein tensor should consist of terms linear in second derivatives or quadratic in first derivatives of the metric. This demand is of little immediate interest to us here, since the other two conditions that Einstein omitted are more revealing. These are:

- the conservation laws (one uses the conservation laws to fix the ratio of the constants C_1 and C_2 in $G_{\mu\nu} = C_1 R_{\mu\nu} + C_2 g_{\mu\nu} R$),
- the Newtonian limit (one uses the Poisson equation to fix the proportionality factor in the constant κ: $\kappa = 8\pi G_{\mathrm{N}}$, with G_{N} Newton's gravitational constant).

Einstein did not include these two constraints in his autobiography, but they were prominent from the Zurich notebook right through to the November 1915 publications. So Einstein failed to mention in 1949 precisely those constraints which he had used heavily when working along the physical strand of his dual method. Could they have been left out on purpose? Even though such an intentional omission may not be very likely, it is fair to observe that quite often an autobiography is not the best source for biographic facts, and this seems to be true in this case too. As Einstein put it himself, "every reminiscence is colored by today's state of affairs, and therefore a deceptive point of view."[85] It appears that his own autobiography suffered from the same bias, as indeed the above passage from the autobiography reads more like a unified field theory manifesto than a historically balanced representation of events. If Einstein had aimed at the latter, his disclaimer was well suited: his omission of the physical strand of his dual method gives an inaccurate picture of the road to the field equations of general relativity. This representation of events is of course understandable in light of the developments of the fall of 1915, but it is unduly one-sided if it is to reflect the full process – starting in 1907, through the Zurich notebook and Entwurf theory years – that led Einstein to his gravity theory.

In the historical analysis that we have presented, one can roughly discern three stages.

- Einstein used a dual method, simultaneously employing physical and mathematical arguments. Working this way he came close to the Einstein equations in 1913, but because of

---

[83] Einstein (1949a), pp. 88–89.
[84] See for example Weinberg (1972), pp. 151–155.
[85] See Einstein (1949a), pp. 2–3.

their conflict with prejudices that were grounded in his network of physical demands, he discarded them.

- Einstein subsequently let the physical arguments prevail over what seemed mathematically the most natural approach and was thus led to the problematic Entwurf theory. He tried to develop the theory further by employing the dual strategy, in particular as he attempted to refine the mathematical structure of the theory.
- When he became convinced of the inadequacy of this theory, he returned to a starting point that was mathematically more natural and rapidly deduced the correct field equations, aided by his enhanced knowledge of the relevant mathematical formalisms, and overcoming problematic physical prejudices along the way.

When looking back in 1949, the guiding role of the dual method that had brought him to the doorstep of the success of 1915 had dropped out of sight. Yet, in reality it had been the interplay of physics and mathematics, in the three phases above, that had given Einstein the conceptual tools that made it possible to take the final steps to complete the theory. When we look at the full development from 1913 until 1915, the ripening physics-first phase of the Entwurf theory had been just as necessary for the final success as a return to the Riemann tensor: it was in this phase that Einstein for instance learnt new variational methods, did his first calculations of the perihelion advancement and thought hard about issues of general covariance. The dual method had produced a theoretical infrastructure that, once reinterpreted from the starting point of the Riemann tensor, quickly yielded the final theory. Jürgen Renn has identified such moments of reinterpretation as "Copernicus" processes: Einstein in November 1915 left the technical machinery and broad outline of the Entwurf theory intact while rearranging basic conceptual and foundational elements. This resembles the thought process of Copernicus, who, according to Renn, "largely kept the deductive machinery of traditional astronomy when changing its basic conceptual structure."[86]

It is likely that, in his own recollection, Einstein gradually came to believe that for a prolonged period he had been led by the nose by faulty abstractions from his physical assumptions. Perhaps he therefore unconsciously failed to mention in his autobiography the above two constraints, or perhaps they dropped from his memory altogether. At the same time, later in life, Einstein did recall positively that he had been saved by the mathematically most natural Ansatz of general covariance. As we will see, these recollections influenced Einstein's engagement with the unified field theory research program; it even directed some of the choices he made in his actual research practice. On the other hand, the recollections themselves, and their presentation, were quite likely shaped just as much by Einstein's commitment to that program.

---

[86] As in Renn (2005b), p. 32; for related ideas, see Janssen and Renn (2007), pp. 910–917.

In the immediate years following the final formulation of general relativity, Einstein was still quite critical of the virtue of formal, mathematical analysis in uncovering the laws of nature. For instance, in 1916 he wrote to Hermann Weyl that Hilbert's "axiomatic method" would fall short for understanding the structure of the electron.[87] Yet he would gradually change his point of view. After all, already in the first paper of the memorable November 1915 series, Einstein had extolled the triumph of Riemannian geometry. Two weeks later he had the correct value for the perihelion shift of Mercury, and another week later, he at last had found *general covariance*, "one of the most exciting [...] moments of my life."[88]

---

[87] "[A]xiomatische Methode," Einstein to H. Weyl, 23 November 1916, Doc. 278, pp. 365–366, on p. 366 in Schulmann *et al.* (1998). Similar early comments on the priority of physics over mathematics are found for example in letters to Theodor Kaluza of 21 April 1919, Doc. 26, pp. 38–40 in Buchwald *et al.* (2004), to Hermann Weyl (5 May 1921, EA 24 091) and to Lorentz (30 June 1921, EA 16 541). See also Corry (1998).

[88] "[...] eine der aufregendsten [...] Zeiten meines Lebens." Einstein to A. Sommerfeld, 28 November 1915, Doc. 153, pp. 206–209, on p. 206 in Schulmann *et al.* (1998).

2

On the method of theoretical physics

Einstein was much engaged with philosophy. His publications on epistemological issues are numerous and some have proven to be important on their own account. Like many of the leading physicists that were his contemporaries – men such as Erwin Schrödinger and Max Planck – he saw engagement with the philosophy of science as part of the intellectual project of physics. Einstein was deeply involved with questions concerning, for example, how scientific creativity works and how abstract thought relates to the actual world.[1]

In Einstein's case it is impossible to disentangle the philosopher from the physicist. Throughout his work there is an evident overlap of the two, as well as a clear interaction between them. As already suggested in the previous chapter, Einstein's physics determined his philosophical outlook, and his philosophy inspired the directions he took in physics. Thus, in striving for a coherent understanding of Einstein's later oeuvre, one needs to address the exchanges between his practices in theoretical physics and his expressed philosophical beliefs – even if the latter were not intended to constitute an elaborated and consistent system.

In this chapter we turn in more detail to Einstein's declared methodological positions. The most important source for these is his 10 June 1933 Herbert Spencer lecture at the University of Oxford, "On the Method of Theoretical Physics." We will study the various ideas laid out in that lecture and try to see how they developed gradually in the years following the discovery of general relativity.

2.1 The change in epistemological outlook

In his 1933 lecture Einstein drew on the historical development of science and how this had influenced ideas about scientific methodology in the past. He derived

---

[1] Philosopher of science Don Howard (2004), section 7, even claims that "Einstein's influence on twentieth-century philosophy of science is comparable to his influence on twentieth-century physics." See also Howard (Forthcoming), p. 8.

his own view in part from his personal experience – most importantly, from the discovery of the general theory of relativity. But he gave other examples as well, including his then current research on spinors and the Dirac equation.

Einstein began by addressing the "eternal antithesis" between the empirical and the rational in the development of a theory. He revered the achievements of the Greeks, who for the first time had come up with a precise logical system, namely Euclid's geometry. According to Einstein, Euclid's geometry could be thought of as a physical science, that is, as dealing with the relations between physical rigid bodies in space. In this perspective it exemplified rather well the function of reason in science: reason is employed in the logical deduction of conclusions – that should correspond to the observed phenomena – from a theory's fundamental concepts and laws. So "the structure of the system is the work of reason; the empirical contents and their mutual relations must find their representation in the conclusions of the theory."[2]

The role of reason was thus quite straightforward. The important and also much debated question that remained was how one arrives at the fundamental concepts and laws of a theory. Einstein lectured in Oxford that Galileo had seen correctly that "pure logical thinking cannot yield us any knowledge of the empirical world; all knowledge of reality starts from experience and ends with it."[3] This statement could easily be taken to imply that as a philosopher, Einstein should be placed in the empiricist tradition. Yet such a conclusion would be too rash, certainly if it is to apply to the time that he spoke these words.

2.1.1 Einstein versus Mach

In the philosophy of science, the debate between realism and empiricism has often taken center stage. Generally, the empiricist tradition has held that theories only have to give an account of what is observable, counting further postulated structure as a means to that end. It has shrunk from the reification of a reality independent of the observed phenomena. Realism, on the other hand, has maintained that the claims of scientific theories are true of a reality independent of us.[4]

The empiricist attitude of Ernst Mach, prominent in the late nineteenth century, implied that one should remain as close as possible to the experimental facts: one should confine oneself to searching out the most economic statements of relations among observations. Broadly speaking, one can say that Einstein moved from Mach's empiricism, earlier in his career, to a strong realist position later on.[5] But at no given moment can one decisively put him in either camp. Even though he was

---

[2] As in Einstein (1933a), on p. 298 in Einstein (1994); "eternal antithesis" on p. 297.
[3] See Einstein (1933a), on p. 297 in Einstein (1994).
[4] An introduction to the realism versus empiricism debate is found in Dieks (1994).
[5] See Holton (1968).

well informed of the antithesis of both schools of thought, at any time one finds elements of one and the other in his thinking. Einstein felt that the creative scientist cannot be too much restricted by the adherence to a particular epistemology. Consequently, the scientist "must [...] appear to the systematic epistemologist as a type of unscrupulous opportunist."[6] Einstein was well aware of his own philosophical opportunism.

Illustrative in this respect are the letters that he exchanged in 1918 with the mathematician-philosopher Eduard Study. Study had published a book with the title *The Realistic World View and the Theory of Space – Geometry, Intuition and Experience.*[7] It discussed the foundations of geometry in the context of the debate on scientific realism. Einstein had read the book and liked it, of which he informed Study. Einstein further wrote to Study that he did not feel comfortable with any of the "isms." His letter reveals his eclectic attitude: on the one hand Einstein found that "Natural science deals with the 'real'," yet "[I] am not a 'realist'."[8] He applauded positivism – a particular brand of empiricism – when it attacked the idea that some concepts are founded in the "a priori." But empiricism was sometimes carried too far, as for instance its campaign against the existence of atoms had shown. Einstein nevertheless believed that Study would immediately qualify his position regarding the relation of geometry to physics as that of an empiricist.

Ernst Mach had opposed atomism from an empiricist position. Atoms do not take part in the human observational experience, thus, said Mach, we cannot ascribe them real existence. For Mach, real existence was reserved for "elements" of sensation. Science is no more than a condensed description of experience, and beyond descriptions, it does not seek explanations. Rather, science seeks economy of description and should therefore dismiss hypotheses and concepts which are not carried by observational elements. This was Mach's principle of economy. It not only inspired his critical stance on atoms, but also his qualification of Newton's absolute space as a "conceptual monstrosity," since it is "purely a thought-thing that cannot be pointed to in experience."[9]

Unsurprisingly, the advent of relativity theory was seen by many as a triumph of the Machist philosophy, and Einstein acknowledged and paid tribute to Mach throughout most of his life.[10] As Gerald Holton has argued, Mach's influence is indicated in Einstein's first paper on the special theory of relativity – the theory that

---

[6] Einstein (1949b), p. 684. Don Howard nevertheless believes that Einstein did develop a coherent philosophy, and actually denies that Einstein started from an empiricist position; for Howard's ideas, see Howard (2004), and Howard (Forthcoming).

[7] Study (1913).

[8] "Ich gebe zu, dass die Naturwissenschaft vom „Realen" handelt und bin doch nicht ein „Realist"." 25 September 1918, Einstein to Study, Doc. 624, pp. 890–892, on. p. 890 in Schulmann *et al.* (1998).

[9] "Begriffsunget[üm] [...] bloss Gedankending, das in der Erfahrung nicht aufgezeicht werden kann." pp. x, 222–223, Mach (1912). For more on Mach's philosophy, see for example Cohen (1970).

[10] See for instance his eulogy for Mach (Einstein, 1916a).

dismissed the superfluous electromagnetic ether. It started with: "We have to take into account that all our judgments in which time plays a part are always judgments of simultaneous events. If for instance I say 'that train arrived here at seven o'clock', I mean something like this: 'The pointing of the small hand of my watch to seven and the arrival of the train are simultaneous events'."[11] One recognizes a sensationist view of measurement and of the concepts of space and time.

In 1909, Mach endorsed the theory of relativity. Indeed, there is a suggestive correspondence between Mach's elements and Einstein's events:[12] the time of an event only becomes a meaningful concept once our consciousness connects to it through sense experience, i.e. when it is subjected to measurement by means of a clock at the same position. The time-coordinate t by itself has no meaning. The special theory of relativity further produced its own philosophical offshoots of empiricism: Einstein's successful analysis of the concept of simultaneity had for instance been an inspiration for the American physicist Percy Bridgman, who formulated an "operationalist" point of view. According to the operationalist philosophy, all scientific concepts should be linked to experimental procedures – science should be cleansed of operationally undefinable terms, as these are devoid of empirical meaning.[13]

However, Holton has further pointed out that Einstein's understanding of experience, of what is to be understood by "empirical fact," grew much broader than Mach's. For Einstein, by 1918, these included the impossibility of a perpetuum mobile, the constancy of the velocity of light, the first law of Newton or the equivalence of inertial and gravitational mass. Experience ranged from direct sensory perceptions to basic laws and notions of physics.[14] These would never have been called facts of experience by Mach.

Mach chose to distance himself again from relativity in 1913, after having been informed of Einstein's first publications on the general theory of relativity. The details of his motivation have remained unknown, as he soon passed away. In fact, his dissociation only became public as late as 1921, when Einstein also learnt of Mach's change of position. Holton believes that Mach had smelled out that

---

[11] "Wir haben zu berücksichtigen, dass alle unsere Urteile, in welchen die Zeit eine Rolle spielt, immer Urteile über *gleichzeitige Ereignisse* sind. Wenn ich z. B. sage: „Jener Zug kommt hier um 7 Uhr an," so heisst dies etwa: „Das Zeigen des kleinen Zeigers meiner Uhr auf 7 und das Ankommen des Zuges sind gleichzeitige Ereignisse."" As in Einstein (1905c), p. 893, translation taken from Holton (1968), on p. 242 in Holton (1988).

[12] As again astutely observed by Holton, pp. 242–243 in Holton (1988). Later, however, Einstein did on occasion distinguish the two; see his letter to Moritz Schlick of 21 May 1917, Doc. 343, pp. 456–457 in Schulmann *et al.* (1998); for commentary, see Howard (2004), sections 2 and 4, Howard (1984), pp. 619–620.

[13] See for example Bridgman (1949). Dennis Dieks argues however that Einstein did not necessarily seek to promulgate an operationalist philosophy himself when he introduced rods and clocks in his definition of simultaneity (Dieks, 2010).

[14] See Holton (1968). The above examples are Holton's, who took them from Einstein's letters to Besso, 28 August 1918, Doc. 607, pp. 864–865 in Schulmann *et al.* (1998), "Erfahrungsthatsache," on its p. 864, and to Ehrenfest, 4 December 1919, Doc. 189, pp. 266–270 in Buchwald *et al.* (2004).

Einstein was much less of a critical empiricist than his philosophy allowed for. Mach reproached relativity as becoming ever more "dogmatic;"[15] this critique might well have been directed at Einstein's acceptance of the equivalence of inertial and gravitational mass as a "fact of experience."

Einstein's later writings contain many realist avowals, completely contrary to Mach's position. For instance, he wrote in 1931 that "the belief in an external world independent of the perceiving subject is the basis of all natural science."[16] At first sight, as said, it seems that he had followed a philosophical pilgrimage, starting from Mach's empiricism and ending in a strong realist position – which became most widely known through his critique of quantum mechanics. Yet things are not that simple, as we can tell, for instance, from the words that Einstein spoke about Galileo in Oxford. Later in life Einstein should nevertheless be thought of as a convinced realist, but sometimes he would nuance that position until a modest empiricist remained: "[I]t is existence and reality one wishes to comprehend. But one shrinks from the use of such words, for one soon gets into difficulties when one has to explain what is really meant by 'reality' and 'comprehend' in such a general statement." What follows puts Einstein right back into the empiricists' camp: "When we strip the statement of its mystical elements we mean that we are seeking for the simplest possible system of thought which will bind together the empirical facts."[17] Einstein did not feel a need to remove the ambivalence and arrange his epistemological position in a systematic fashion. But then, the traditional realism versus empiricism dichotomy may even have appeared unnatural to him. As he wrote to Eduard Study: "If one cleanses two arbitrary 'isms' of all rubbish, they become each other's equivalent."[18]

2.1.2 Intuition versus induction

Einstein's gradual drifting away from Mach's philosophy appears to have gone hand in hand with a remarkable change of position on how scientific creativity works. In his 1949 autobiography, after granting Mach to have inspired his work on relativity, Einstein criticized him for not having placed "in the correct light the essentially constructive and speculative nature of thought."[19] A year before, he had

---

[15] From the preface in Mach (1921), quoted on p. 248 in Holton (1988). Gereon Wolters (1984) has claimed however that Mach had not dissociated himself from relativity at all, but rather that the "anti-relativist" Mach was the result of forgery by his son, Ludwig Mach, who supposedly meddled with his father's posthumous text. Holton (1993) disagrees.

[16] As in Einstein (1931), on p. 291 in Einstein (1994).

[17] Taken from Fine (1996), p. 106. According to Arthur Fine, Einstein's words are from a lecture at Columbia University, published in 1934. However, he points out that John Stachel cites this source as a lecture at the University of California, Los Angeles in 1932, see Stachel (2002), p. 373. The documentation of the Einstein Archive for this entry (EA 2 110) is not specific enough to decide either way.

[18] "Wenn man zwei beliebige „ismusse" von allem Unrat säubert, dann werden sie einander gleich." 25 September 1918, Einstein to Study, Doc. 624, pp. 890–892, on p. 890 in Schulmann *et al.* (1998).

[19] From p. 21 in Einstein (1949a).

written to Michele Besso:

[Mach] did not recognize the freely constructive element in the formation of concepts. In a way he thought that theories arise through *discoveries* and not through *inventions*. He even went so far that he regarded "sensations" not only as the material which has to be investigated, but, as it were, as the building blocks of the real world; thereby he believed he could overcome the difference between psychology and physics. If he had drawn the full consequences, he would have had to reject not only atomism but also the idea of a physical reality.[20]

Einstein's Spencer lecture at Oxford also contained a number of positions that were at odds with Mach's philosophy on this issue. On that occasion Einstein stressed the logical contingency of induction from the phenomena and ultimately claimed its impracticability. Mach's search for economical relations among the elements of experience, however, strongly reflects the method of induction.

Einstein underpinned his dismissal of induction by referring to the demise of Newtonian theory, as this had been effected by the general theory of relativity. The Newtonian principles are quite different from the principles of general relativity, yet both correspond with experience to a large extent; this proved that abstraction of "the basic concepts and postulates of mechanics from elementary experience is doomed to failure."[21]

What does one generally understand by induction? Usually, induction is described as the logical inference of general statements from individual facts. Nineteenth century inductivists such as John Stuart Mill (in this context Einstein also referred to Mach[22]) believed that if one is given suitable experiences, then by following a certain number of inductive recipes one will be led to discover the relevant laws. Einstein's critique should be seen in the context of this contention, but it was equally directed at broader understandings of induction as ampliative argumentation on the basis of experience. We should further keep in mind that he conceived of "experience" in a rather general way – as we just saw, fundamental concepts and laws of physics could be included.

Einstein had already expressed himself dismissive of the role of induction on a number of occasions preceding the Oxford lecture. On Christmas day of 1919 he had published in the *Berliner Tageblatt* the essay "Induction and deduction in physics." Quite similar to how he had chosen his words five years earlier in his

---

[20] "[Mach] hat das freie konstruktive Element in der Begriffsbildung verkannt. Er meinte gewissermassen, dass Theorien durch *Entdeckung* und nicht durch *Erfindung* entstehen. Er ging sogar so weit, dass er die „Empfindungen" nicht nur als zu erfassendes Material, sondern gewissermassen als die Bausteine der wirklichen Welt ansah; er glaubte so die Differenz zwischen Psychologie und Physik überwindern zu können. Wenn er ganz konsequent gewesen wäre, hätte er nicht nur den Atomismus sondern die Idee einer physikalischen Realität ablehnen müssen." 6 January 1948, Einstein to Besso, pp. 390–392, on p. 391 in Einstein and Besso (1972), translation as on pp. 249–250 in Holton (1988); emphasis as in the original.

[21] See Einstein (1933a), p. 300 in Einstein (1994).

[22] See Einstein (1936a), p. 333 in Einstein (1994). John Stuart Mill's epistemology is generally identified as "inductivism"; according to Mill all scientific inquiry – not only in the context of discovery, but also in the context of the justification of a theory – uses exclusively inductive methods.

inaugural address at the Academy, it began with an outline of the scientific method that reminds one of the dual strategy:

> The simplest idea about the development of science is that it follows the inductive method. Individual facts are chosen and grouped in such a way that the law, which connects them, becomes evident. By grouping these laws more general ones can be derived until a more or less homogeneous system would have been created for this set of individual facts. Starting from these generalizations, the retrospective mind could then, inversely, arrive back at the individual facts by pure reasoning.[23]

The 1919 essay in the *Berliner Tageblatt*, however, took a different turn than the 1914 address at the Prussian Academy. In his inaugural address, Einstein had contended that "inductive physics has questions for deductive physics and vice versa." But by 1919 induction had fallen into disgrace: "The truly great advances in our understanding of nature originated in a manner almost diametrically opposed to induction."[24] Induction was replaced by "intuition":

> The intuitive grasp of the essentials of a large complex of facts leads the scientist to the postulation of a hypothetical basic law, or several such basic laws. [...] [The researcher] does not find his system of ideas in a methodical, inductive way; rather, he snuggles up to the facts by intuitive selection among the conceivable theories that are based upon axioms.[25]

In 1914 Einstein had also contended that there was no "learnable, systematically applicable method" that leads to the sought after principles; yet advances were still expected to come from a joint application of "inductive physics" and "deductive physics." Indeed, he had told his Academy audience that "eavesdrop[ping] on nature" was to be done by "recognizing in larger sets of experiential facts certain general traits."[26] In 1919, however, a perspective on scientific practice that involved induction was dismissed; instead, the researcher's "intuitive grasp" had become the essential creative agent.

It is tempting to point to Einstein's formulation of general relativity as an essential ingredient of an explanation of his changed attitude towards induction. It would

---

[23] "Die einfachste Vorstellung, die man sich von der Entstehung einer Erfahrungswissenschaft bilden kann, ist die nach der induktiven Methode. Einzeltatsachen werden so gewählt und gruppiert, dass der gesetzmässige Zusammenhang zwischen denselben klar hervortritt. Durch Gruppierung dieser Gesetzmässigkeiten lassen sich wieder allgemeinere Gesetzmässigkeiten erzielen, bis ein mehr oder weniger einheitliches System zu der vorhandenen Menge der Einzeltatsachen geschaffen wäre von der Art, dass der rückschauende Geist aus den so gewonnenen letzten Verallgemeinerungen auf umgekehrtem, rein gedanklichem Wege wieder zu den Einzeltatsachen gelangen könnte." As in Einstein (1919c).

[24] "Die wahrhaft grossen Fortschritte der Naturerkenntnis sind auf einem der Induktion fast diamentral entgegengesetzten Wege entstanden," Einstein (1919c).

[25] "Intuitive Erfassung des Wesentlichen eines grossen Tatsachenkomplexes führt den Forscher zur Aufstellung eines hypothetischen Grundgesetzes oder mehrerer solcher. [...] [Der Forscher] gelangt nicht auf methodischem, induktivem Wege zu seinem Gedankensysteme, sondern er schmiegt sich den Tatsachen an durch intuitive Auswahl unter den denkbaren, auf Axiomen beruhenden Theorien" (Einstein, 1919c). A similar emphasis on the role of intuition is found in Einstein (1918).

[26] As in Einstein (1914b); see note 54 in Chapter 1.

of course be a stretch to identify the arguments that produced the Entwurf theory as typically inductive as opposed to those that gave general relativity, yet there is a certain resonance between Einstein's change of position and the experience of 1915. Arguments leading him in the direction of the Entwurf theory were typically ampliative, bottom-up arguments, anchored in familiar physics; examples are Einstein's inference to the Newton metric (1.15) or the generalization of the familiar Poisson equation to the core gravitational term in the Entwurf field equation, $\gamma_I^{\mu\nu}$ from equation (1.3). As we saw at the end of Chapter 1, Einstein was greatly surprised when he learnt that the Poisson equation was in fact "incorrect" at the lowest, easiest observable order; generalizing upon its basis had thus been too precarious. Similarly, general relativity with its firmer role for the equivalence principle can easily be imagined as inspiring Einstein to put more emphasis on the creative merit of "the intuitive grasp" of a "hypothetical basic law." Seen in this light, the experience of 1915 then does provide a natural perspective from which to understand Einstein's changed assessment of induction.

Einstein did not change his critical stance regarding induction after 1919. The Einstein Archives contain a 1935 letter from Einstein to the philosopher Karl Popper and in that letter he agreed, among other things, with Popper's critique of the "inductive method" contained in the latter's *Logic of Scientific Discovery*:[27]

Your book has pleased me very much in many ways: rejection of the "inductive method" from an epistemological standpoint. Also falsifiability as the crucial property of a theory (of reality). [...] You have further defended your positions really well and astutely.[28]

In his first chapter, Popper discussed "the problem of induction," i.e. he addressed the question: are inductive inferences logically justified? Popper argued that they are not. He found that inductive inferences can only be justified if they can be compared to a formal logical principle of induction, a principle that would in all generality encompass how one should go from singular facts to universal laws. This principle of induction would make it possible to put inductive inferences in a logically acceptable form; inductive statements could then be reduced to pure logical truths, such as tautologies. Citing David Hume, Popper claimed that the introduction of a principle of induction led to an infinite regress: such a principle would be a universal statement, inferred from past experience by induction. In other words, the justification of the principle of induction would require inductive inferences, and to justify these one would need an inductive principle of higher order, and so on.

[27] See Popper (1935).
[28] "Ihr Buch hat mir in vieler Hinsicht sehr gefallen: Ablehnung der „induktiven Methode" vom erkenntnistheoretischen Standpunkte. Auch die Falsifizierbarkeit als entscheidende Eigenschaft einer (Wirklichkeits) Theorie. [...] Sie haben Ihre Positionen auch wirklich gut und scharfsinnig verteidigt." 15 June 1935, Einstein to Popper, EA 19–124.

Like Einstein, Popper pointed to the logical contingency of inductive arguments. As only experience can decide upon the acceptance or rejection of a theory, he next asked the question how we can ever know whether the natural laws of science are in fact true. He concluded that we cannot: we can only test their falsity. The "empirical method" exposes a particular scientific system to falsification. The aim is to select the "fittest" theory, the one that is strongest in surviving tests. Popper claimed that this method of falsification does not need to presuppose the logical validity of inductive inferences, but only uses "tautological transformations of deductive logic, whose validity is not in dispute."[29]

As the above letter illustrated, Einstein was really taken by the whole of Popper's book, including the falsification thesis, and he offered to bring the book to the attention of his colleagues. That Einstein might have liked Popper's idea of falsification can be no surprise, as he had made a similar point in his 1919 essay in the *Berliner Tageblatt*.[30] It had been written a little over a month after Einstein had acquired world fame by the announcement of the observational confirmation of the gravitational bending of light-rays. Perhaps to dampen the less subtle public expressions of acclaim – the same *Berliner Tageblatt* had reported the 1919 eclipse results by exclaiming that "a highest truth, beyond Galileo and Newton, beyond Kant" had been unveiled by "an oracular saying from the depth of the skies"[31] – he pointed to the reservations with which any scientific theory should be approached:

A theory can thus be recognized as erroneous if there is a logical error in its deductions, or as inadequate if a fact is not in agreement with its consequences. But the *truth* of a theory can never be proven. For one never knows that even in the future no experience will be encountered which contradicts its consequences; and still other systems of thought are always conceivable which are capable of joining together the same given facts.[32]

The falsification thesis tried to sidestep the problem of induction. Thus settling the epistemic status of scientific theories may have resolved Popper's worries, but

---

[29] Quoted from Popper (1959), p. 42

[30] Since Einstein's position there and Popper's later work exhibited similarities, John Stachel, as editor of the Einstein Papers, wrote to Popper asking whether he knew of and had been inspired by Einstein's 1919 piece. Upon receiving the article from Stachel in 1984, Popper said that he had never seen it before, but was struck by its contents. Popper further explained to Stachel what did lead him to reject induction. This was the overthrow of Newtonian physics by Einstein's theory: "Einstein's theory clashed logically with Newton's while containing Newton's as an approximation. [This] had a decisive influence on my views on science: it established the invalidity of induction which plays such a role in Newton. It shows that any established theory may be merely a first approximation." 15 March 1984, Popper to Stachel, as in Adam (2000), p. 23; for more on Einstein's influence on Popper, see Hacohen (2000), pp. 94–96. Another early statement in which Einstein expressed falsificationist ideas is found in Einstein (1922), p. 429.

[31] "[…], dass eine letzte Wahrheit entschleierbar war über Galileo und Newton, über Kant hinaus, bestätigt durch einen Orakelspruch aus der Tiefe des Himmels," Moszkowski (1919).

[32] "Eine Theorie kann also wohl als unrichtig erkannt werden, wenn in ihren Deduktionen ein logischer Fehler ist, oder als unzutreffend, wenn eine Tatsache mit einer ihrer Folgerungen nicht in Einklang ist. Niemals aber kann die *Wahrheit* einer Theorie erwiesen werden. Denn niemals weiss man, dass auch in Zukunft keine Erfahrung bekannt werden wird, die ihren Folgerungen widerspricht; und stets sind noch andere Gedankensysteme denkbar, welche imstande sind, dieselben gegebenen Tatsachen zu verknüpfen." As in Einstein (1919c).

Einstein lectured in Oxford that experience is "the alpha and omega of all our knowledge of reality," the beginning and end of all knowledge: he still needed to understand how he was able to formulate theories, if not by induction. By 1919, his answer had become, as we have seen, that the foremost determining factor was "the intuitive view of the researcher."[33]

2.1.3 Mathematical naturalness instead of induction and a priori syntheses

Historically, Immanuel Kant's philosophy had indicated a way out of Hume's inductive skepticism. In Popper's words, Kant took the principle of induction to be a priori valid; that is, as a way out of the infinite regress, Kant had claimed that it is possible to make valid a priori synthetic judgments. He justified this by pointing to the human intuition (that is, "Anschauung"). Kant's point of view in the end implied that there would have to be statements about the world which are necessarily true; this, however, was unacceptable to Einstein.

Let us try to clarify this issue. Suppose events of type A have always been observed to be followed by events of type B. Then, one would like to claim universally that A causes B. Yet, experience can never show that such a law is strictly universal. The judgment that A causes B must therefore be grounded in an additional a priori faculty of knowledge: our faculty of understanding. When a causal law is grounded in the a priori, we can universally – and must necessarily – assert that events of type A are followed by events of type B. Contra Hume, Kant argued that there are such necessary and more than merely inductive laws. Kant found examples of these in the principles of Newtonian mechanics, i.e., the law of conservation of matter or the equality of action and reaction. These are synthetic a priori principles: human understanding guarantees the a priori truth of these principles. Our understanding can further elevate inductively established rules to necessary and universal laws. Kant illustrated this by comparing Kepler's rules for the motions of the planets with Newton's laws of gravitation: where Kepler's rules are just a grouping of data, the motions of the planets are necessarily prescribed by Newton's laws of gravitation.[34]

In the Oxford lecture, Einstein used the same example to dismiss Kant's ideas as he had used to dismiss induction: since the Newtonian principles had been replaced by a very different set of (relativistic) principles, one cannot but conclude that such fundamental principles have a fictitious character. The most important dividing line

---

[33] "[D]en intuitiven Bild des Forschers," Einstein (1919c).

[34] For a contemporary Anglo-American perspective on Kant, see for example Friedman (1992). The above aims primarily to reflect Einstein's understanding of Kant; see in that respect Einstein (1944), p. 23 in Einstein (1994); see also Beller (2000), p. 89.

between Einstein and Kant was thus the different status that they attributed to the fundamental axioms and laws of nature; to Einstein, these were in principle entirely tentative.

As Moritz Schlick proclaimed in public, and as Einstein agreed to Schlick in private,[35] relativity theory announced the demise of the Kantian system. The Kantian philosophy, according to Schlick, "may itself be regarded as a product of the Newtonian doctrine of nature."[36] Kant had held that we have a priori knowledge – knowledge not derived from any experience – that Euclidean geometry is true in the world we perceive. It is a precondition, a necessary form of our intuition, under which we organize the perception of spatial relations between objects. In the same way, arithmetic is a precondition for the perception of time. In turn, Euclidean geometry and arithmetic seemed to justify the observation that we have a capacity to make correct a priori synthetic claims: we know for sure, without having to count apples, that $1017 + 2913 = 3930$. The great appeal of Kant's system had been that it warranted the feeling that geometry, arithmetic and the fundamental principles of Newtonian kinematics were too certain to be based merely on generalization from experience.[37] The special theory of relativity and its four-dimensional spacetime continuum, however, dismissed Kant's a priori three-dimensional Euclidean space. Schlick argued that one needed to distinguish intuitional, psychological spaces and non-intuitional, physical space; Kant had not distinguished the two clearly enough,[38] and one senses that Schlick believed that Kant had consequently put Euclidean space on the pedestal of the a priori.

In a letter, written just three weeks after the formulation of the general theory, Einstein applauded Schlick's 1915 critique of Kant:

[T]ruly masterful is your position to the doctrine of Kant and his followers. The trust in the "apodictic certainty" of "synthetic a priori judgments" is already heavily undermined if one realizes the invalidity of even just one of these judgments.[39]

Einstein did not fail to call attention to the weaknesses of the Kantian theory. In an article in the literary journal *Die neue Rundschau* (edited by Rudolf Kayser,

---

[35] See Einstein to Schlick, 13 December 1915, Doc. 165, pp. 220–221 in Schulmann *et al.* (1998). For more on their correspondence, see Howard (1984), Hentschel (1986).

[36] Taken from Schlick (1915), citation on p. 153 of Schlick (1979).

[37] See Glymour (1999), pp. 109–110.

[38] See Schlick (1917), p. 55.

[39] "[W]ahrhaft meisterhaft [ist] ihr Verhältnis zur Lehre Kants und seiner Nachfolger. Das Vertrauen auf die „apodiktische Gewissheit" der „synthetischen Urteile a priori" wird schwer erschüttert durch die Erkenntnis der Ungültigkeit auch nur eines einzigen dieser Urteile." 14 December 1915, Einstein to M. Schlick, 14 December 1915, Doc. 165, pp. 220–221 in Schulmann *et al.* (1998). In this letter, Einstein continued with praise for both Mach and Hume, and said that the latter in particular may have been indispensable in the conception of special relativity. He also agreed with Schlick's observation that relativity is closely related to positivism, but that it does not presuppose positivism.

Einstein's son-in-law) he even came close to ridiculing Kant's conception of geometry,[40] and in letters sent to fellow physicists he not only rejected Kant, but also embraced Hume.[41]

The 1921 paper *Geometry and experience* implicitly contrasted new ways of understanding geometry to Kant's with the simplest of examples; Einstein raised the question of how we are to interpret the axiom of Euclidean geometry that just one straight line connects two points:

> The older interpretation: everyone knows what a straight line is, and what a point is. Whether this knowledge springs from an ability of the human mind or from experience, from some cooperation of the two or from some other source, is not for the mathematician to decide. He leaves the question to the philosopher. Being based upon this knowledge, which precedes all mathematics, the axiom stated above is, like all other axioms, self-evident, that is, it is the expression of a part of this *a priori* knowledge.
>
> The more modern interpretation: geometry treats of objects which are denoted by the words straight line, point, etc. No knowledge or intuition of these objects is assumed but only the validity of the axioms, such as the one stated above, which are to be taken in a purely formal sense, i.e. as void of all content of intuition or experience. These axioms are free creations of the human mind. All other propositions of geometry are logical inferences from the axioms (which are to be taken in the nominalistic sense only).[42]

Einstein felt that the fundamental concepts and principles of geometry and physics cannot be grounded in some a priori fashion. Instead, he emphasized that they are "free creations of the human mind." To be sure, this freedom is not the same freedom as that which a writer of fiction may enjoy. It is rather like that "of a man solving a well-designed word puzzle. He may [...] propose any word as the solution, but there is only *one* word which really solves the puzzle in all its parts." But a freedom it is; to believe that the abstract notions of science could somehow be

---

[40] "The attempt to lift geometry from the murky realm of the empirical, led, unnoticed, to a conceptual change that is somewhat analogous to the promotion to gods of the revered heroes of antiquity." ("Das Bestreben, die ganze Geometrie aus den trüben Sphären des Empirischen herauszuheben, führte nun unvermerkt zu einer geistigen Umstellung, welche der Beförderung verehrter Helden der Vorzeit zu Göttern einigermassen analog ist.") As in Einstein (1925b), pp. 16–17.

[41] See Einstein to Ehrenfest, 24 October 1916, Doc. 269, pp. 346–347 in Schulmann *et al.* (1998); Einstein to Born, 29 June 1918, Doc. 575, pp. 818–819 in Schulmann *et al.* (1998). For Einstein's opinion of Hume versus Kant, see for example also Einstein (1949a), p. 13.

[42] "Ältere Interpretation. Jeder weiss, was eine Gerade ist und was ein Punkt ist. Ob dies Wissen aus einem Vermögen des menschlichen Geistes oder aus der Erfahrung, aus einem Zusammenwirken beider oder sonstwoher stammt, braucht der Mathematiker nicht zu entscheiden, sondern überlässt diese Entscheidung dem Philosophen. Gestützt auf diese vor aller Mathematik gegebene Kenntnis ist das genannte Axiom (sowie alle anderen Axiome) evident, d.h. es ist der Ausdruck für einen Teil dieser Kenntnis a priori. – Neuere Interpretation. Die Geometrie handelt von Gegenständen, die mit den Worten Gerade, Punkt usw. bezeichnet werden. Irgendeine Kenntnis oder Anschauung wird von diesen Gegenständen nicht vorausgesetzt, sondern nur die Gültigkeit jener ebenfalls rein formal, d.h. losgelöst von jedem Anschauungs- und Erlebnisinhalte aufzufassenden Axiome, von denen das genannte ein Beispiel ist. Diese Axiome sind freie Schöpfungen des menschlichen Geistes. Alle anderen geometrischen Sätze sind logische Folgerungen aus den (nur nominalistisch aufzufassenden) Axiomen." As in Einstein (1921a), pp. 4–5 (an earlier, shorter version of this lecture appeared in the *Sitzungsberichte der Königlich Preußischen Akademie der Wissenschaften*, pp. 123–130, 1921).

a priori conditioned and therefore unalterable was erroneous and even constituted a "serious danger to the progress of science."[43]

The defective Entwurf theory had highlighted that fundamental equations cannot be found by straightfowardly expanding on known physics, and relativity had further taught Einstein that primary concepts cannot be a priori justified in our faculty to understand the world around us. So without the certitude of syntheses grounded in the a priori, and without the method of induction, the intuitive creation of the axioms of physics might very well be a very speculative and precarious leap. But then, how can a creative physicist ever know if he is on the right track?

Einstein asked this question in his Oxford lecture, and on that occasion he presented a quite revealing answer. Appealing to "our experience hitherto" – echoing the experience with general relativity – he suggested that the right track is charted by the observation that "nature is the realization of the simplest conceivable mathematical ideas." Einstein, in 1933, was convinced that "we can discover by purely mathematical constructions the concepts, and the laws connecting them, which furnish the key to the understanding of the natural phenomena." According to his assessment, "the creative principle resides in mathematics." Experience of course remained the sole criterion of the physical utility of a mathematical construction. But it was the proven creative merit of striving after mathematical naturalness and simplicity that gave Einstein the confidence that "pure thought can grasp reality."[44]

The free choice of theoretical principles was thus shaped by the criterion of mathematical simplicity, the practical validity of which had been established by Einstein's experiences with general relativity. In the 1919 article in the *Berliner Tageblatt* there was not yet any mention that a researcher's intuitions were guided by the idea that nature is the realization of the simplest possible mathematical laws. We have pointed out before that this element of Einstein's later epistemology was not yet present in the years that immediately followed the conception of general relativity[45] (and we will return to the issue when we discuss his relation to experiment). The heuristic of mathematical simplicity, however, gradually gained prominence in Einstein's scientific practice in the next decade, until it was formally announced and firmly anchored as his true guiding principle in the Spencer lecture at Oxford; in the years following 1933 we do not see much change in Einstein's

---

[43] As in Einstein (1936a), on p. 324 and p. 329 in Einstein (1994). Einstein's emphasis on the freedom to choose the fundamental principles of physics resonated with the conventionalist perspective, which maintained that there are always alternative theories conceivable that explain equally well all available empirical evidence, and that from a logical point of view, the choice among these theories is thus a matter of convention. Einstein's relation to conventionalism is discussed in detail in Howard (1984, 1990a).

[44] As in Einstein (1933a), on p. 300 in Einstein (1994). See also Norton (2000).

[45] Abraham Pais (1982), p. 325, draws attention to a letter from Einstein to Felix Klein, dated 15 December 1917, in which Einstein reproached the latter for overrating formal points of view. Einstein even proclaimed that they "fail almost always as heuristic aids" ("[S]ie versagen fast stets als heuristische Hilfsmittel"), Doc. 408, pp. 569–570, on p. 569 in Schulmann *et al.* (1998).

methodological beliefs. Finally, let us note that the idea that nature adheres to the simplest mathematical representation available of course also echoes Mach's economic principle.[46]

As to Einstein's position with regard to Kant, some nuances also need to be made. On a number of occasions Einstein actually expressed himself quite appreciative of Kant's ideas, and some aspects of Einstein's thought did rather resemble the Kantian philosophy.[47] Both for instance emphasized the virtue of striving for unity in science. In 1949, Einstein would even state that his philosophy is in fact "distinct from that of Kant only by the fact that we do not conceive of the categories as unalterable"[48] – rejecting only Kant's a priori necessary truths. But even on this issue, Einstein transgressed his own convictions to some degree: de facto, he would never depart from classical field theory and in the context of quantum mechanics, for example, he did venture that possibly this would one day dispense of the spacetime continuum, yet all the same deemed such a program an "attempt to breathe in empty space."[49]

2.2 Principle and constructive theories

The article in the *Berliner Tageblatt* was not the only piece that Einstein wrote following the successful British eclipse expeditions: a similar expository note was published in the London *Times* on 28 November 1919, only three weeks after the eclipse results had been announced. This article gave an outline of relativity theory for a lay audience, and made reverential bows to Isaac Newton – these were likely included to please a British public that had grown resentful of German science during World War I. Einstein again presented an epistemological discussion of the nature of physical theories. He made a distinction between "principle" and "constructive" theories, and discussed the development of relativity theory in the context of this distinction.

According to Einstein, constructive theories attempt to "build up a picture of the more complex phenomena out of the materials of a relatively simple formal scheme from which they start out."[50] An example was the kinetic theory that reduced the mechanical and thermal properties of gases to movements of molecules. In 1919 Einstein seemed particularly to appreciate these kinds of theories: he argued that "understanding" a group of natural phenomena usually meant that one has found a constructive theory to describe them with.

[46] On the relation between Einstein's "simplicity" and Mach's economic principle, see Einstein (1950), p. 13.

[47] For Einstein's positive commentary on Kant, see for example Einstein (1936a), p. 320 in Einstein (1994); for a reading of Einstein's philosophical ideas as (neo-)Kantian philosophy, see Beller (2000).

[48] As in Einstein (1949b), p. 674.

[49] As in Einstein (1936a), on p. 352 in Einstein (1994).

[50] See Einstein (1919b); cited from p. 248 in Einstein (1994).

Einstein contended that principle theories do not start out from hypothetical constructions, but rather from empirically discovered principles. These principles form criteria that systems described by the theory have to satisfy. "Thus the science of thermodynamics seeks by analytical means to deduce necessary conditions, which separate events have to satisfy, from the universally experienced fact that perpetual motion is impossible." Principle theories, Einstein held, "employ the analytic, not the synthetic method."[51] Both constructive and principle theories had their advantages: constructive theories were more complete, adaptable and clear, whereas principle theories had greater "logical perfection and security of the foundations."[52]

In Einstein's discussion of relativity in *The Times*, the general principle of relativity took center stage: it was supported by the "fact of experience" that the inertial and gravitational masses of a body are equal, which in turn had inspired the principle of equivalence. Einstein identified relativity theory as a principle theory; he thus qualified relativity as structured as an "analytic" top-down theory instead of a "synthetic" bottom-up theory. However, the strategy employed by Einstein as he was formulating general relativity seems poorly represented by a principle approach alone. The dual strategy that we discussed in Chapter 1 would be better captured by a combination of both analytic and synthetic approaches. Einstein's discussion of the theory in *The Times*, however, indicated a clear deductive process: he suggested that a "consistent following up" of the equivalence principle had in the end justified hopes that a general theory of relativity would give the laws of gravity.[53] Again we see Einstein, when looking back, failing to represent the dual path that had initially led him to the Entwurf theory and that was instrumental in deepening his understanding of the conceptual issues underlying a theory of general relativity. Also in the language of the principle–constructive dichotomy, as in his autobiography, Einstein did not display the relevance of his bottom-up approaches to the gravity problem prominently enough, and instead emphasized the positive role of a principle–deductive manner of research.

Einstein's distinction between principle and constructive theories is sometimes presented as a central tenet of his philosophy,[54] but the number of occasions that he expressed himself in these categories is really quite limited. For instance, one does not clearly recognize the principle–constructive distinction in the 1933 Herbert Spencer lecture. Einstein scholar John Stachel in fact believes that the distinction became blurred later in Einstein's career,[55] and we agree with his observation. In

---

[51] See Einstein (1919b); cited from pp. 248–249 in Einstein (1994). On this distinction in relation to Einstein's philosophy, see also Howard (2005).

[52] See Einstein (1919b); cited from p. 248 in Einstein (1994).

[53] See Einstein (1919b); cited from pp. 251–252 in Einstein (1994).

[54] See for example Howard (2004), section 6.

[55] Stachel (1991), on p. 409 in Stachel (2002).

Einstein's later writings, progress in "understanding" is attributed, for instance, to the unification of theories, with the end-result resembling a principle rather than a constructive theory.

2.3 Einstein's methodological schema

To end this chapter we would like to illustrate Einstein's methodological ideas once more by turning to a letter that he wrote to his lifelong friend Maurice Solovine on 7 May 1952. Einstein and Solovine had both been members of the "Academia Olympia," the make-believe academy in which Einstein had debated science and philosophy in his early professional life in Bern. In 1952 they were again corresponding about epistemological issues and Solovine had become puzzled by one of Einstein's positions. In his latest reply – "Dear Solo! In your letter you give me a spanking on the behind [...], but, probably I expressed myself badly. Schematically I see the matter as follows;..."[56] – Einstein gave Solovine a remarkably illuminating exposition of his later methodological conviction, mirroring ideas of his Spencer lecture of two decades before. He drew a diagram in which the function of intuition, deduction and experience were clearly laid out.

The letter to Solovine and its diagram were brought to our attention by an instructive article by Gerald Holton, and we will start this section by closely following Holton's treatment.[57] We have reproduced Einstein's diagram here in Figure 2.1 and have redrawn it in Figure 2.2. In the diagram, Einstein depicted the various stages involved in theory construction; it captured a circular process that began and ended at a line at the bottom, indicated with a letter "E". Einstein explained:

1. The E (experiences) are given to us.[58]

The line E represented the "manifold of immediate (sense) experiences." As Holton has taught us, these were not necessarily confined to the sense experiences proper, if we take into account Einstein's positions regarding the laws of nature: the constancy of the velocity of light, the equivalence of inertial and gravitational mass, or even the Poisson equation from Newtonian gravitation theory may be included.[59]

Out of the "manifold of experience" ascended a curved and searching arrow. It reached the apex of the structure, where Einstein located "A," the theory's system of axioms:

---

[56] "Lieber Solo! In Ihrem Brief geben Sie mir [...] auf den Popo [...]; wahrscheinlich habe ich mich slecht ausgedrückt. Ich sehe die Sache schematisch so; [...]" Einstein to M. Solovine, 7 May 1952, pp. 118–121 in Einstein (1956), quoted on pp. 118, 120.

[57] See Holton (1998b); quotations in English from Einstein's letter are also as in Holton's paper.

[58] "1. Die E (Erlebnisse) sind uns gegeben."

[59] See Holton (1998b), p. 32. In a review of Émile Meyerson's book "La déduction relativiste" Einstein attributed a similar wide ranging conception of experience to Meyerson and did not distance himself from this conception, see Einstein (1928c), p. 161.

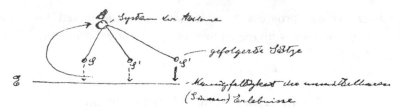

Figure 2.1. Einstein's methodological schema, as drawn in his letter to Maurice Solovine of 7 May 1952 (by permission of the Albert Einstein Archives, Hebrew University, Jerusalem).

2. *A* are the axioms from which we draw consequences. Psychologically the *A* are based upon the *E*. There is however no logical path from the *E* to the *A*, but only an intuitive (psychological) connection, which is always "subject to revocation."[60]

He again stressed the contingent and intuitive character of this leap from the phenomena to the *A*, the physical principles. This was why the formulation of theories was so uncertain: there is no secure way of abstracting a theory's concepts from experience *E*.[61] At the time of his 1914 inaugural lecture, Einstein might still have allowed more straight, "inductive" arrows, but they were absent in his 1952 letter to Solovine; instead, he drew a delicate and fallible curve, expressing the unsure nature of the "free creations of the human mind."

According to Einstein's epistemology, very many different axiom systems *A* could in principle be arrived at in the top of his diagram. This of course greatly complicated the theorist's task. Yet, there was a helpful heuristic that could serve as a guide when making the jump to the *A*: for Einstein, by 1933, the route was charted by the belief "that nature is the realization of the simplest conceivable mathematical ideas." Obviously, the demand for mathematical simplicity also came in at the end of the process, when one had to decide whether to favor one theory over another, when both can account for roughly the same set of observed facts.

We come to deduction. In the letter to Solovine, Einstein wrote:

3. From *A*, *by a logical path*, particular assertions are deduced – deductions which lay claim to being right.[62]

---

[60] "2. *A* sind die Axiome, aus denen wir Folgerungen ziehen. Psychologisch beruhen die *A* auf *E*. Es gibt aber keinen logischen Weg von den *E* zu *A*, sondern nur einen intuitiven (psychologischen) Zusammenhang, der immer „auf Widerruf" ist."

[61] Holton (1998b), p. 35, points out that perhaps this discontinuity is represented by the small gap that Einstein left between the line *E* below and the arc that rises to *A* in his drawing.

[62] "3. Aus *A* werden *auf logischen Wegen* Einzel-Aussagen *S* abgeleitet, welche Ableitungen den Anspruch auf Richtigkeit erheben können."

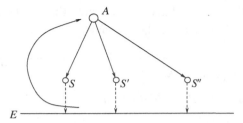

Figure 2.2. Einstein's schema. Next to the letters *E*, *A* and *S*, he wrote: *E* "Mannig-faltigkeit der unmittelbaren (Sinnes-)Erlebnisse" [manifold of immediate (sense) experiences], *A* "System der Axiome," *S* "gefolgerte Sätze" [deduced theorems].

Following the creative leap, one deduced: if the set of axioms *A* applied, assertions *S*, *S'*, *S''*, and so on, must necessarily ensue. Finally, experience entered again:

4. The *S* are related to the *E* (testing against experience). Carefully considered this procedure also belongs to the extra-logical (intuitive) sphere, because the relations between concepts appearing in *S* and the experiences *E* are not of a logical nature. This relation of the *S* to the *E* is, however (pragmatically), far less uncertain than the relation of the *A* to the *E*.[63]

Einstein now assessed whether the structure, part postulate and part inference, would fit with observation and experiment; that is, whether the conclusions and predictions *S* matched with experience *E*. If that was the case, one could put greater trust in the entire cycle. But one could never be certain that a theory is correct: there is always the possibility that additional observations would prove it wrong.

Until here we have followed Einstein's letter, and Holton's analysis, of what we will call the "Solovine schema." A natural question that next comes to mind is how the route to the general theory of relativity is reflected in the schema, given that route's central role in Einstein's epistemology. As we know, he had claimed in his "Autobiographical notes" that the gravitational equations were only found "through the discovery of a logically simple mathematical condition," that is, "the invariance concerning the continuous transformation group." The demand of general covariance "determine[d] the equations completely or almost completely."[64] This suggests a top-down course of events, and it thus seems that Einstein's description of his route to general relativity in his autobiography fits the schema that he drew in 1952 well. When trying to capture Einstein's description, and expanding on it in obvious ways, one may give the following epistemological sketch of the

---

[63] "4. Die *S* werden mit den *E* in Beziehung gebracht (Prüfung an der Erfahrung). Diese Prozedur gehört genau betrachtet ebenfalls der extra-logischen (intuitiven) Sphäre an, weil die Beziehung der in den *S* auftretenden Begriffe zu den Erlebnissen *E* nicht logischer Natur sind. Diese Beziehung der *S* zu den *E* ist aber (pragmatisch) viel weniger unsicher als die Beziehung der *A* zu den *E*."
[64] As in Einstein (1949a), p. 89.

route to the general theory of relativity – a sketch that we believe Einstein would have subscribed to in essence in his later years:

The relevant "fact of experience" was:

E *equivalence of inertial and gravitational mass*

One now freely jumped to a set of axioms, guided by the conviction that the laws of nature necessarily abide by the simplest mathematical formulation:

A *equivalence principle*
  general covariance
  Riemannian geometry

Through a deductive process would follow:

S *field equations*
  Schwarzschild solution
  geodesic particle trajectories

These reverted to experience again:

E *perihelion motion of Mercury*
  bending of light-rays
  gravitational redshift

But we have already seen that Einstein's autobiography is historically inaccurate, as it practically ignored the stage of the Entwurf theory. After completion of the final theory, Einstein's struggles with the Entwurf theory gradually sunk into oblivion, and if at all, he would only mention them in passing.[65] We have seen on the other hand that his actual research in the years 1913 through 1915 followed a two-pronged strategy, a strategy that we termed his dual method and that was based on the double application of "bottom-up" and "top-down" arguments. Can we recognize the dual method, in a natural way, in Einstein's schema?

We find that we cannot, even though his diagram has both creative arrows rising out of E and deductive arrows that point in the opposite direction. The trouble lies along the physics-first, bottom-up strand: it does not seem to be properly included

---

[65] As we have seen, Einstein did not usually mention the Entwurf stage. An exception is his 1922 Kyoto address, entitled "How I created the theory of relativity," but he only briefly referred to it in that lecture; see Abiko (2000), p. 16. In his 1933 "Notes on the origin of the general theory of relativity," the Entwurf stage is merely presented as a time in which he could not see how the final field equations "could be used in physics" as he believed them to be at odds with experience; he further briefly mentioned the hole argument. Both prejudices had hampered Einstein: they were "errors of thought which cost me two years of excessively hard work" (Einstein, 1933b), on pp. 317–318 in Einstein (1994).

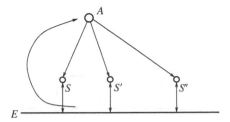

Figure 2.3. Sketch of the dual method. The essential difference from Einstein's original picture drawn for Solovine are arrows pointing from E right to the S: arrows representing physical arguments that initially led to the Entwurf field equations and played an important role in the construction of general relativity.

in Einstein's diagram. The precarious, curved arrow that connects E to A should not be the only arrow that emanates from experience, if the diagram is to reflect accurately the dual method.

There is no problem with the initial "intuitive" step, from E: *equivalence of inertial and gravitational mass* to A: *equivalence principle* and *general covariance*. In the Zurich notebook, we have subsequently seen Einstein deduce from the Riemann tensor, moving to the S: *field equations*. The top-down approach to the field equations, started off by an initial creative and intuitive leap, seems well reflected in the Solovine diagram.

But this was not the only way of approaching the field equations S. The other route went straight up out of the manifold E. With Einstein's broad understanding of experience in mind, it should be appropriate to place *the Newtonian limit* of gravity – the Poisson equation, and for example its solar system solutions – in E. Then, the Zurich notebook generalizations to dynamical field equations that began from the Newtonian limit, as in equation (1.3), should be seen as direct attempts to construct field equations S starting with "experience" E, without a detour through the A.

Such attempts would be captured naturally by arrows pointing directly from the E to the S in the diagram. Likewise, Einstein had started with familiar physics when combining Newtonian theory with special relativity into the "Newton metric" (1.15), again an essential ingredient in both the Zurich notebook and in the publications of the Entwurf theory. However, these steps are not easily recognized in the diagram sent to Solovine: Figure 2.2 has no arrows pointing upwards from the E to the S. The dual method would look something like Figure 2.3, which does include arrows coming out of the experiences E to the field equations S.

The lack of arrows pointing from E to the S in Figure 2.2 becomes problematic if we want to interpret the stage in which Einstein upheld the Entwurf theory by using his diagram. During that period the upward swinging curved arrow, connecting E to

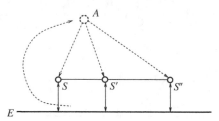

Figure 2.4. The Entwurf theory was predominantly based on bottom-up inferences that were more direct generalizations of familiar physics.

A, would have looked rather dim, if present at all, owing to the theory's problematic relation to general covariance. Einstein's confidence in the Entwurf field equations S was based primarily on arguments that came out of Newtonian physics, that is on arrows pointing up from E to the S; see Figure 2.4. If Einstein had generally covariant field equations – even though the hole argument had initially told him that they were physically uninteresting – and had been able to show that the conservation laws restricted them to the Entwurf form, he would have known how to draw the arrows from A to S. Instead, he could only suppose such connections to be there, despite for instance his attempts at a unique Lagrangian formulation.

Einstein never deduced the Entwurf field equations S from a system of axioms A, but nonetheless had a strong belief in the correctness of the Entwurf theory. That belief was grounded primarily in arguments that came from his physical background knowledge, anchored in E; for instance in the fact that the Entwurf theory promptly reduced to the familiar Poisson equation in the appropriate limit, or that it contained the desired Newton metric.

We have seen that during the reign of the Entwurf, Einstein tried to show that the symmetries of the conservation laws – which restricted the coordinate system – and the residual symmetries of the field equations were the same. Such a proof would give the scheme some solidity (indicated in Figure 2.4 by the horizontal lines) and, if realized, would also offer hope of some compliance with the axiom of general covariance. However, when Einstein found that the Entwurf field equations were not even covariant upon a transformation to a rotating system, his confidence in the possibility that there could be arrows pointing from the equivalence principle A to the Entwurf S crumbled, and he abandoned the theory.

Given the above, it is clear that the method that Einstein actually used from 1913 to 1915 to close in on the field equations is not easily recognized in his 1952 diagram. The diagram does not reflect the positive contributions of the learning phase of the Entwurf theory in which bottom-up physical arguments held the upper hand over a top-down mathematically more natural alternative, just as this stage had been ignored in the autobiography.

When looking at the course of events in the fall of 1915, such omissions cannot be a surprise: in November of 1915, Einstein finally let the deductive arguments derived from the mathematical Riemann tensor prevail over his physical presumptions and promptly found the correct field equations. As pointed out before, this aspect would eventually come to dominate his recollections. In 1938, for instance, he wrote to his fellow physicist Cornelius Lanczos:

Coming from skeptical empiricism of somewhat the kind of Mach's, I was made, by the problem of gravitation, into a believing rationalist, that is, into someone who seeks the only trustworthy source of truth in mathematical simplicity.[66]

The route to general relativity made Einstein reshape his methodological beliefs. He took great stock in his revised ideas; on many occasions, as we have seen, he presented his prized theory as if it truly had been found by following the recipe of the Solovine schema 2.2. Quite likely, if any of his readers would have reminded him of the positive contributions of the physics-first Entwurf phase, he would have replied that general relativity nevertheless really *ought* to have been found by the epistemological route captured by his figure. As the following chapters will argue, the Solovine schema reflected well his practice in unified field theory.

---

[66] "Vom skeptischen Empirismus etwa Mach'scher Art herkommend hat das Gravitationsproblem mich zu einem gläubigen Rationalisten gemacht, d.h. zu einem, der die einzige zuverlässige Quelle der Wahrheit in der mathematischen Einfachheit sucht." 24 January 1938, Einstein to C. Lanczos, EA 15–268; also quoted in Holton (1988), p. 259.

3

Unification and field theory

There is an element of the Oxford lecture that we have not yet addressed in much detail: Einstein's claim that "the grand object of all theory" is to unify – to make the fundamental concepts and laws "as simple and as few in number as possible."[1] Likewise, he had told Cornelius Lanczos in 1938 that "the physically true is logically simple, that is, it has unity in its foundation" – even if the reverse does not necessarily hold ("the logically simple does not, of course, have to be physically true."[2])

Unification had always been an important aspect of Einstein's work – we need only to think of the general theory of relativity, where he unified gravitation with special relativity – and it played a prominent role in his epistemology. Some ten papers by him carried the word "Einheitliche" in their title. These were all papers on unified field theories: theories that attempted to unify in a single mathematical, preferably geometrical scheme the gravitational and electromagnetic fields. We wish to outline the role of unification in Einstein's philosophy here, followed by a brief introduction to the unified field theory program.

3.1 Unification: motivation and implementation

In his autobiography Einstein gave two criteria for a successful theory: firstly, of course, it should not contradict empirical facts. Secondly, the theory should display "inner perfection," which was characterized by its "naturalness," usually mathematically construed, and "logical simplicity."[3] A theory should certainly not be a patchwork: "in view of its task in ordering and surveying sensory experiences, [it] should show as much unity and parsimony as possible."[4] Einstein admitted that

[1] As in Einstein (1933a), on p. 298 in Einstein (1994).
[2] "Das logisch Einfache braucht zwar nicht physikalisch wahr zu sein, aber das physikalisch Wahre ist logisch einfach, d.h. einheitlich in seiner Grundlage." 24 January 1938, Einstein to C. Lanczos, EA 15-268.
[3] Einstein (1949a), pp. 21, 23; see also Holton (1998b) and Howard (2004), section 3.
[4] See Einstein (1944), on p. 25 in Einstein (1994).

such qualities cannot easily be assessed in an objective, un-intuitive manner, and one will not find a final definition in his writings of what he meant by terms such as logical simplicity or mathematical naturalness. Nevertheless, Einstein held that there is usually agreement "among the augurs" in judging the inner perfection of a theory. One of the obvious roles of unification was of course to enhance a theory's "inner perfection."

This criterion for the quality of a theory was grounded in a strong Platonist conviction: ultimately, all attempts at unification were grounded in "the belief [...] that the structure of the existing world is perfectly harmonious."[5] This belief also emerges naturally when comparing two of Einstein's positions discussed earlier: his realism and his falsificationalism. If science aims to give true stories of what the world independent of our senses is like, yet if we can never know whether a scientific theory is actually true, then the scientist's work might very well be an entirely vain enterprize. The idealism that followed from Einstein's Platonist conviction, however, made the contradiction less pregnant: unifying theories, and increasing their mathematical naturalness, brings us closer and closer to the harmony that encompasses the "real in all its depth."[6]

Einstein held that the unification of theories gives scientific progress: "A theory is the more impressive the greater the simplicity of its premises is, the more different kinds of things it relates, and the more extended is its area of applicability."[7] When referring back to the Solovine schema, we see that progress can be achieved by connecting the axioms A of theories that were beforehand unconnected, and thus bring more and more phenomena – more and more matches between the S and E – under a simpler axiom system; "thus the story goes on until we have arrived at a system of the greatest conceivable unity, and of the greatest possible poverty of concepts of the logical foundations."[8] In this process, say, two Solovine-schemas come together and merge by having their axiom systems A_1 and A_2 derived from an encompassing system A' (see Figure 3.1). The latter may very well have been arrived at by exclusively theoretical or even mathematical study.

Gerald Holton has identified in the work of many scientists the role played by thematic presuppositions, or "themata."[9] These guide the directions taken in theory construction: they filter and mold the possible leaps from E and are instrumental

---

[5] "[...] der Glaube, [...] dass das Seiende in seiner Struktur von vollendeter Harmonie sei." Taken from an unpublished and untitled manuscript in the Einstein Archive, entry no. EA 2-110, dated to 1931. Victor F. Lenzen (1949) also identified a Platonist trait in Einstein's philosophy in his article on "Einstein's theory of knowledge" in the Schilpp volume; Einstein (1949b), p. 683, qualified the article as "convincing and correct."

[6] As in Einstein (1933a), on p. 301 in Einstein (1994). Mara Beller (2000), pp. 90–91, said that Einstein attached no meaning to the "truth" of a theory, when dissociated from a criterion of systematic unity in capturing the totality of physical experience; presupposing a methodological principle of unity and the presence of a systematic unity of nature would further resonate with Kant's ideas.

[7] As in Einstein (1949a), p. 33.

[8] As in Einstein (1936a), on p. 323 in Einstein (1994).

[9] See Holton (1998a,b).

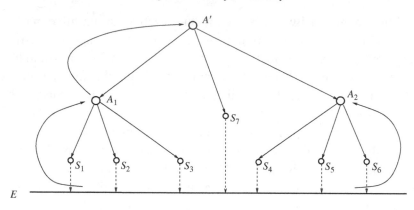

Figure 3.1. The unification of two theories: as the leap from A_1 and A_2 suggests, this is largely a theoretical exercise. The unified structure would hopefully cover a wider range of phenomena than the two theories separately. Unification reduces the contingency of the entire structure, as one unified theory is logically simpler than multiple disjunct theories. Even if no new phenomena can be incorporated, it is still desirable: "Of two theories that represent the same complex of phenomena compatible with observation, the theory that is grounded on the lesser number of logically independent hypotheses (axioms) is to be preferred." ("Von zwei Theorien, die den selben Komplex von Phänomenen im Einklang mit den Beobachtungen darstellen, ist diejenige zu bevorzugen, welche auf weniger logisch voneinander unabhängige Hypothesen (Axiome) gegründet ist." Einstein to Julian H. Bonfante, 15 October 1944, EA 6-116.)

in determining what A one arrives at. In Einstein's case, one readily recognizes as one of his themata the emphasis put on mathematical naturalness, logical simplicity and unification. One easily points to the cultural milieu of nineteenth century Germany as a relevant source for such maxims. No lesser figures than Alexander von Humboldt and Johann Wolfgang von Goethe, the latter in particular through his Faust figure, had emphasized the striving of science for simplification on the one hand and the ultimate unity of nature on the other.[10] While the origins of these ideas are located in romantic natural philosophy of the early nineteenth century, Einstein likely absorbed them for the first time through the works of popular science writers like Ludwig Büchner and Aaron Bernstein (who had in fact considered themselves critical of the natural philosophers). Bernstein believed that all of nature's forces were different manifestations of one fundamental force and claimed that the "next significant level of natural science will be one in which the unity of force can be demonstrated."[11] Einstein had read Bernstein as a young boy, although it is not known precisely which of Bernstein's numerous writings he had picked

[10] See Holton (1998a), pp. 21–22; Neffe (2007), p. 56.
[11] Aaron Bernstein, *Naturwissenschaftliche Volksbücher* (1870), Berlin: Franz Duncker, vol. 5, p. 143, as quoted in Gregory (2000), on p. 33.

up. However, it is certain that he had read Ludwig Büchner's *Kraft und Stoff*, and Büchner found that "simplicity is known to be the hallmark of truth."[12]

Bernstein and Büchner were widely read and they influenced an entire generation. Such popular notions of unity and simplicity should likely have inspired Einstein's efforts at unification; these cultural resources evidently make up a relevant part of the broader context of Einstein's work. Yet, he himself pointed to other sources for inspiration – in particular the experience of general relativity – and most importantly, they are not particular to him, precisely because of their broad influence. This suggests that by themselves, they are insufficient for understanding why Einstein set out on, and persisted in, his later unification attempts, as opposed to others of his generation who did not. But this does not mean that the reverberations of German Romanticism or late nineteenth century scientific popularizers should be denied a shaping role; on the contrary, they would have resonated well with his personal experiences in the discovery of general relativity. Indeed, with both combined, Einstein's engagement with the unified field theory program and the philosophy of his 1933 Spencer lecture can hardly seem surprising.

When turning to Einstein's practice in physics, an important question that arises is: how did mathematical naturalness, logical simplicity (or mathematical simplicity) and unification relate to one another, and how did they manifest themselves in Einstein's work? We first give an example of a situation in which they were fairly equivalent: in relativity, one no longer dealt with a separate inertial and gravitational mass, unlike in Newtonian mechanics. With the unification of these concepts one had removed an "internal asymmetry," which was felt to be "unnatural."[13] Undoubtedly, mechanics had become logically simpler as well. In a similar vein, Einstein sought to re-derive the geodesic equation from the gravitational field equations,[14] and most importantly, he wanted to describe particles as non-singular solutions of a set of field equations. This would unify the field and particle concepts, which had been two disjoint objects since Lorentz's formulation of electrodynamics. This problem was a central theme in Einstein's oeuvre, particularly in his later years, and we will return to it on several occasions.[15]

---

[12] "Einfachheit ist bekanntlich das Kennzeichen der Wahrheit," Büchner (1888, p. 13). See Gregory (2000) for more on Bernstein and Büchner.

[13] See p. 31 in Einstein (1949a).

[14] The most elaborate paper on this subject is Einstein *et al.* (1938); for a recent contribution to the historiography of this problem, see Kennefick (2005).

[15] Arnold Sommerfeld wrote to Einstein in 1937: "I heard from Weyl [...] that you tirelessly seek out the elementary particles in the most hidden folds and branchings of the general world differential equations. If I will see the day when the electron will no longer be a stranger in electrodynamics, as you once said around 1908??" ("Ich hörte von Weyl, [...], dass Sie den Elementarpartikeln unermüdlich nachspüren, in den verborgenen Falten und Verzweigungen der allgemeinen Welt-Differential-Gleichungen. Ob ich es noch erleben werde, dass das Elektron nicht mehr ein Fremdling in der Elektrodynamik ist, wie Sie etwa 1908 einmal gesagt haben??") 30 December 1937, pp. 117–118 in Einstein and Sommerfeld (1968), on p. 118; Sommerfeld was likely referring to p. 192 in Einstein (1909a).

In terms of field theory, Einstein understood by unification often the same as physicists regard it to be today – namely as the coming together in one Lagrangian of, what were originally, disparate fields. A theory is "more nearly perfect" and "simpler" when it is invariant under a larger group of transformations.[16] Simplicity, according to Einstein, is also a good criterion once one already knows what group the theory should be invariant under: in the case of the general theory of relativity, there are a number of field equations imaginable that would exhibit general covariance; it turns out that the simplest equations are the correct equations.[17]

Einstein came to believe that general relativity had a number of shortcomings. Firstly, it appeared not to contain non-singular particle solutions (he gave a mathematical proof of this point in an article that appeared in Argentina in 1941).[18] It also had not attained a high enough degree of unification:

[T]his theory [has] in my opinion little intrinsic likelihood. The field variables $g_{\mu\nu}$ and A_μ do not correspond to a *unified* conception for the structure of the continuum. Furthermore, the left and right hand side of the field equations [...] are juxtaposed in a logically arbitrary way (that is, without any formal constraint).[19]

Einstein found that the introduction of the stress-energy tensor in the general relativistic field equation could only be considered a provisional completion of the theory. "It seems unnatural to introduce such concepts as density and velocity [as in $T_{ik} = \rho u_i u_k$] as concepts of the same depth as the tensor g_{ik}. But even the independent introduction of the electromagnetic field has generally been felt to be unsatisfactory."[20] The logical arbitrariness with which the Maxwell term had been added to the Einstein–Hilbert action deprived the theory of some of its mathematical, esthetic appeal: "it is similar to a building, one wing of which is made of fine marble (left part of the equation), but the other wing of which is built of low grade wood (right side of equation)."[21]

Arbitrariness, instead of simplicity, was a manifestation of the lack of unity of a theory. In the end, such a shortcoming was due to our lack of understanding: "[We mean] by simplicity [...], that the system contains as few independent assumptions or axioms as possible; for the totality of logically independent axioms stands for the 'un-understood remainder'." Einstein characterized his own strivings in physics particularly as a search for unity: "My own efforts first and foremost concern the

---

[16] As on p. 77 in Einstein (1949a).

[17] See p. 69 in Einstein (1949a).

[18] See Einstein (1941); for more on the article and its proof, see van Dongen (2002); Galvagno and Giribet (2005).

[19] "[...] diese Theorie [besitzt] nach meiner Ansicht wenig innere Wahrscheinlichkeit. Die Feldvariabeln $g_{\mu\nu}$ und A_μ entsprechen keiner *einheitlichen* Konzeption für die Struktur des Kontinuums. Auch die linke und rechte Seite der Feldgleichungen [...] sind in logisch willkürlicher Weise (d.h. ohne formalen Zwang) neben einandergestellt." Emphasis as in original (manuscript, EA 2-111, "Über den gegenwärtigen Stand der allgemeinen Relativitätstheorie," 1930).

[20] "On the present state of the theory of relativity," 1947, EA 2-126, manuscript.

[21] As in Einstein (1936a), on p. 342 in Einstein (1994).

logical unity in physics." After all, "one wants to *understand* the existing, real world."[22]

As we would expect on the basis of our discussion in Chapter 2, by the 1930s Einstein believed that such true understanding could never be attained by inductively grazing off the phenomena. The quest for a logically simple, unified theory was at odds with "the so-called 'phenomenological' physics": a physics that "makes as much use as possible of concepts which are close to experience but, for this reason, has to give up, to a large extent, unity in the foundations." This phenomenological conception of physics coincided with the "theories of knowledge [of] St. Mill and E. Mach" – theories of knowledge that indeed believed that the laws of physics "could be obtained purely inductively from experience."[23] Phenomenological physics was something very different from unification physics.

According to Einstein's epistemology, unification enhanced the explanatory power of a theory – the ability to describe or derive a variety of phenomena with a restricted set of laws. To philosophers of science, a link between "unification" and "understanding" or "explanation" will be no surprise as it has often been argued for (just as arguments that oppose this link have been made)[24] – in Einstein's particular case, the link cohered closely with the nuances of his views on realism, as the following passage from an article in *Scientific American* suggests:

I believe that every true theorist is a kind of tamed metaphysicist, no matter how pure a "positivist" he may fancy himself. The metaphysicist believes that the logically simple is also the real. The tamed metaphysicist believes that not all that is logically simple is embodied in experienced reality, but that the totality of all sensory experience can be "comprehended" on the basis of a conceptual system built on premises of great simplicity.[25]

A "passion for comprehension" thus led Einstein to the "striving toward unification and simplification." He conceded that a skeptic might find that he had fallen for a "miracle creed." Nevertheless, he held that this creed had borne many fruits in the history of science; in the same article Einstein pointed to the ancient atomists, who had been vindicated, and, as we have now come to expect, to his own general theory of relativity as examples of the miracle creed's success.[26]

---

[22] "[Wir meinen] unter Einfachheit [...], dass das System möglichst wenige voneinander unabhängige Voraussetzungen bezw. Axiome enthalte; denn der Inbegriff logisch voneinander unabhängiger Axiome bedeutet den „Rest des nicht Begriffenen". [...] Dies eigene Streben galt ganz vorwiegend der logischen Einheitlichkeit in der Physik. [...] [M]an will das Seiende, das Wirkliche *begreifen*" (EA 2-110, untitled manuscript, 1931; emphasis as in the original).

[23] See Einstein (1936a), on pp. 332–333 in Einstein (1994).

[24] For an overview of philosophies that establish this link, and for a critique of those positions, see Morrison (2000); see also Friedman (1974).

[25] As in Einstein (1950), p. 13; this reference was brought to our attention by an interesting article by Tilman Sauer (Forthcoming).

[26] See Einstein (1950), pp. 13, 15.

3.2 Unified field theory

Einstein's conviction that by unifying we enhance our understanding found its clearest expression in his attempts to unify the gravitational and electromagnetic fields. These attempts at "unified field theories" began to be taken up in Einstein's research in the late 1910s and increasingly preoccupied his work thereafter. The first prominent theory that is today generally identified as a unified field theory was formulated by the mathematical physicist Hermann Weyl in 1918. In 1921 two other important proposals appeared in print: the theory of Theodor Kaluza – extensively studied by Einstein, and we will discuss his work on this proposal in more detail later – and another theory due to Sir Arthur Stanley Eddington.[27] A fourth notable attempt at a unified field theory from the 1920s, based on a "teleparallel" geometry, was of Einstein's own making.[28] In these unified theories, not only the gravitational field, but also the electromagnetic field preferably attained a purely geometric origin – just as the gravitational field had found a geometric formulation in general relativity. This was to be done in such a way that both fields emanated from a single mathematical structure, from which, hopefully, the quantum nature of matter could also be derived.

Weyl's 1918 paper showed that the gauge invariance of electrodynamics was a similar sort of symmetry as the coordinate invariance of gravitation theory. Essentially, Weyl had replaced the connection $\Gamma^{\alpha}_{\mu\nu}$ from relativity by the following, more general expression:

$$\hat{\Gamma}^{\alpha}_{\mu\nu} = \frac{1}{2}g^{\alpha\beta}[(\partial_{\mu}g_{\beta\nu} + \partial_{\nu}g_{\beta\mu} - \partial_{\beta}g_{\mu\nu}) + (g_{\beta\mu}A_{\nu} + g_{\beta\nu}A_{\mu} - g_{\mu\nu}A_{\beta})]$$

$$= \Gamma^{\alpha}_{\mu\nu} + \frac{1}{2}g^{\alpha\beta}(g_{\beta\mu}A_{\nu} + g_{\beta\nu}A_{\mu} - g_{\mu\nu}A_{\beta}). \tag{3.1}$$

With this connection he had incorporated in the geometry gauge transformations on the electromagnetic gauge field A_{μ} as scale transformations on the metric. In his theory, rescaling the metric produced a gauge transformation on A_{μ}:

$$g_{\mu\nu} \rightarrow e^{\Lambda(x)}g_{\mu\nu}, \qquad A_{\mu} \rightarrow A_{\mu} - \partial_{\mu}\Lambda(x), \tag{3.2}$$

with $\Lambda(x)$ an arbitrary function.

The geometric difference from the general theory of relativity followed from comparing the parallel transport of a vector in both theories. In Riemannian geometry the parallel transport of a vector around a closed curve need not be integrable,

---

[27] For historical reviews of the early unified field theory program, see Bergia (1993), Vizgin (1994), Goldstein and Ritter (2003), Goenner (2004, 2005); for a review with a particular focus on Einstein, see Sauer (Forthcoming), and Pais (1982), pp. 325–354.

[28] Mathematicians had however earlier come up with similar ideas; for more on the history of this theory, see in particular Sauer (2006).

i.e. the direction of a vector can change under parallel transport. Its magnitude, however, remains unchanged. This was no longer the case in Weyl's geometry. In fact, Weyl introduced and motivated his unified field theory in his original publication exactly by dismissing this axiom of Riemannian geometry.[29]

Weyl's theory originally created quite a stir. The brash young physicist Wolfgang Pauli immediately set out to see how it compared to the results from relativity – finding that the perihelion shift of Mercury and the gravitational bending of light carried over to the new theory. He also attempted to find a particle solution, but this attempt failed; the equations attained such a degree of complexity that Pauli did not succeed in integrating them. Subsequently, he took a critical position to Weyl's theory.[30]

Pauli's doubts found their way to an argument with which he opposed the very core assumptions of the nascent unified field theory program. Unified field theories were intended to yield non-singular solutions to their field equations that could serve as a model for the quantum electron. Such – to be found – particle solutions presuppose that one can make sense of the electric field in the interior of a particle, Pauli argued. But, the field strength is defined as the force acting on a test-particle, and since there are no test-particles smaller than the electron, the field in the interior of such a solution would be unobservable by definition, and thus "fictitious and without physical meaning."[31]

Pauli did not keep his reservations to himself. At the 1920 "Naturforscherversammlung" in Bad Nauheim he directly addressed Einstein with his, essentially operationalist, objection.[32] He wanted to know whether Einstein agreed to a need for more radical approaches:

I would like to ask prof. Einstein if he agrees that one can only expect a solution of the problem of matter by a modification of our ideas about space (and perhaps also time) and electric fields in the sense of atomism or if he [...] is of the opinion that one should hold on to the rudiments of the continuum theories.[33]

Einstein, in his reply, evaded giving a direct answer. He would not decide either way and made a subtle epistemological remark, concerning the situation when a scientific concept or theory is in opposition with nature. He made a distinction

---

[29] See Weyl (1918), on p. 25 of the English translation contained in O'Raifeartaigh (1997).

[30] On Pauli's engagement with Weyl's theory, see Vizgin (1994), pp. 104–111.

[31] "[E]ine physikalisch inhaltslose Fiktion" (Pauli, 1921, p. 775), p. 206 in Pauli (1958).

[32] Relativity theory was hotly debated at this meeting, and Pauli's critical stance towards Einstein should not be misconstrued as part of the anti-relativity attitude of some quarters of German physics. For more on this meeting, see for example Goenner (1993), Rowe (2006), van Dongen (2007c).

[33] "Ich möchte [...] Herrn Prof. Einstein fragen, ob er der Auffassung zustimmt, dass man die Lösung des Problems der Materie nur von einer Modifikation unserer Vorstellungen vom Raum (vielleicht auch von der Zeit) und vom elektrischen Feld im Sinne des Atomismus erwarten darf, oder ob er [...] die Ansicht vertritt, dass man an den Grundlagen der Kontinuumstheorien festhalten muss." In: "Vorträge und Diskussionen von der 86. Naturforscherversammlung in Nauheim vom 19. bis 25. September 1920," *Physikalische Zeitschrift*, **21**, 1920, pp. 649–666, on p. 650.

between a procedure where a concept is dropped from a physical theory, and a procedure where the system of arrangement of concepts to events is replaced by a more complicated one, that still refers to the same concepts.[34] The concept at hand was the familiar continuous field, and although Einstein did not yet speak out a full commitment in Bad Nauheim, he had in fact already opted for the second approach. It is important to observe that Einstein would continue his research by retaining the familiar field concepts and type of partial differential equations customary in the general theory of relativity.

In 1919, before the Bad Nauheim meeting, Einstein had already ventured that for the stability of possible particle solutions the gravitational field might be required, and he considered altering the gravitational field equations (1.25): Einstein tailored their left-hand side (by replacing the factor of 1/2 by a factor of 1/4) such that the curvature scalar could act as a pressure term that would stabilize the interior of a particle.[35]

Pauli decided against such methods. A year after the Bad Nauheim meeting, he wrote in his elaborate encyclopedia article on general relativity that "[i]t is the aim of all continuum theories to derive the atomic nature of electricity from the property that the differential equations expressing the physical laws have only a discrete number of solutions which are everywhere regular, static and spherically symmetric." He, however, was of the opinion that "the existence of atomicity, in itself so simple and basic, should also be interpreted in a simple and elementary manner and should not [. . .] appear as a trick in analysis."[36]

Pauli's characterization of the aim of the unified field theory program would prove to be quite accurate. Einstein would try to find the type of particle solutions as outlined by Pauli in most of the field theories that he continued to study. However, in 1920 Pauli – still only 20 years old – was convinced that a different route should be followed: "new elements which are foreign to the continuum concept of the field will have to be added to the basic structure of the theories developed so far, before one can arrive at a satisfactory solution of the problem of matter."[37]

When Pauli objected to the implicit assumptions of the unified field theory program by pointing to the fictitious character of field strengths inside elementary

---

[34] See "Vorträge und Diskussionen von der 86. Naturforscherversammlung in Nauheim vom 19. bis 25. September 1920," *Physikalische Zeitschrift*, **21**, 1920, pp. 649–666, on pp. 650–651.

[35] See Einstein (1919a).

[36] "Das Ziel aller Kontinuumstheorien ist, den Atomismus der Elektrizität darauf zurückzuführen, dass die Differentialgleichungen, welche die Naturgesetze ausdrücken, nur eine diskrete Zahl von überall regulären, statischen und kugelsymmetrischen Lösungen haben [. . .]. [M]an wird [. . .] verlangen müssen, dass die an sich so einfache und grundlegende Tatsache des Atomismus auch einfach und elementar von der Theorie zu deuten ist und nicht [. . .] als ein Kunststück der Analysis erscheint," (Pauli, 1921, p. 774); p. 205 in Pauli (1958).

[37] "[Z]u den Grundlagen der bisher aufgestellten Theorien [müssen] erst neue, der Kontinuumsauffassung des Feldes fremde Elemente hinzukommen, damit man zu einer befriedigenden Lösung des Problems der Materie gelangt," (Pauli, 1921, p. 775); p. 206 in Pauli (1958).

particles, his argument echoed Einstein's operationalist emphasis on rods and clocks in the 1905 definition of simultaneity. Pauli had not convinced Einstein, but the latter did resort to the rods and clocks of special relativity to point out a shortcoming of Hermann Weyl's theory. Since in that theory the magnitude of a vector can change along a time-like worldline, there is no way of defining a standard clock: the rate at which a clock would tick at a certain point in spacetime would depend on the history of that clock. As the periodicities of atoms define clocks, this meant that atoms of the same element at a certain point in spacetime could have different spectra, which evidently did not agree with observation. Despite this damning criticism, Weyl did not immediately dismiss his theory and elaborated it for a few more years. Eventually, however, he no longer held out great hopes for it, or even the unified field program on the whole. He did return to gauge symmetries, as we will soon see.[38]

Einstein extensively studied a number of unified field theories in the decade following 1918. In this period he worked out most thoroughly his own teleparallel theory. Like Weyl's theory, this theory was first introduced by outlining a possible extension of the axioms of Riemannian geometry.[39] Both theories used the mathematics of four-dimensional affine spaces. Such geometries had been classified by the mathematician Élie Cartan in a paper from 1923.[40] Cartan's classification involved three quantities:

$$\Omega_\mu^\nu = -R_{\mu\alpha\beta}^\nu \, (\mathrm{d}x^\alpha \wedge \mathrm{d}x^\beta)$$
$$\Omega^\nu = -T_{\alpha\beta}^\nu \, (\mathrm{d}x^\alpha \wedge \mathrm{d}x^\beta) \tag{3.3}$$
$$\Omega = -R_{\mu\alpha\beta}^\mu \, (\mathrm{d}x^\alpha \wedge \mathrm{d}x^\beta)$$

where $\mathrm{d}x^\alpha \wedge \mathrm{d}x^\beta$ is the area of an infinitesimal, closed loop. The $T_{\alpha\beta}^\nu$ is the spacetime "torsion," which is non-zero if the geometry has a non-symmetric connection: $T_{\alpha\beta}^\nu = \frac{1}{2}(\tilde{\Gamma}_{\alpha\beta}^\nu - \tilde{\Gamma}_{\beta\alpha}^\nu)$.

For the geometries referred to so far, Cartan's classification gives the following:

$$\Omega_\mu^\nu = 0, \quad \Omega^\nu = 0, \quad \Omega = 0, \quad \text{Euclidean geometry;}$$
$$\Omega_\mu^\nu \neq 0, \quad \Omega^\nu = 0, \quad \Omega = 0, \quad \text{Riemannian geometry;} \tag{3.4}$$
$$\Omega_\mu^\nu \neq 0, \quad \Omega^\nu = 0, \quad \Omega \neq 0, \quad \text{Weyl's geometry.}$$

[38] Einstein's criticism of Weyl's theory is contained in an addendum to Weyl's article Weyl (1918), on p. 478; extensive commentary on Einstein's note is found in Janssen *et al.* (2002), p. 62. For historical discussion of Weyl's theory, see also Vizgin (1994), pp. 71–112, Goenner (2004), section 4.1, Ryckman (2005), chapter 4, and the various essays in Scholz (2001).

[39] Einstein's first papers on the teleparallel theory are Einstein (1928a,b); for discussion of his theory see for example Vizgin (1994), pp. 234–257, Goldstein and Ritter (2003), pp. 120–133, Goenner (2004), section 6.4 and Sauer (2006).

[40] See Cartan (1923); for a discussion of Cartan's work in the context of unified field theories, see also Tonnelat (1965). For the classification of theories as given here, see Vizgin (1994), pp. 184–185, 235.

Einstein's teleparallel geometry can be characterized by

$$\Omega^{\nu}_{\mu} = 0, \quad \Omega^{\nu} \neq 0, \quad \Omega = 0, \quad \text{teleparallel geometry.} \qquad (3.5)$$

In this geometry both the directions and lengths of vectors remain unchanged if one parallel transports them around a closed loop, or along any curve; thus, for example, one can compare the length and direction of vectors across finite distances and, unlike in the general theory of relativity, there is no Riemannian curvature. The geometric interpretation of the non-vanishing torsion is that it precludes the usual construction of an infinitesimal parallelogram by parallel transport. That is, if both vectors *OA* and *OB* originate in a point *O*, and one parallel transports *OA* along *OB* yielding a line element *BO'*, and *OB* along *OA* giving *AO''*, the points *O'* and *O''* are not the same – the torsion is a measure of the distance separating them. By introducing torsion one has generalized the familiar Euclidean axioms of geometry, as had been done earlier in the case of Riemannian and Weyl's geometry.

Einstein was unaware of Cartan's work, and apparently created his theory without any substantial interaction with mathematicians.[41] Yet, rethinking the geometric axioms was an approach from which Einstein had great expectations:

The aspirations of theorists are aimed at finding natural generalizations or completions of the Riemannian geometry that contain more concepts than that geometry, in the hope to arrive at a logical construction that unifies all physical field concepts in one single point of view.[42]

The teleparallel geometry in principle contained enough fields to describe both gravity and electromagnetism. The fundamental quantity in Einstein's new theory was indeed to be the torsion:

Because the tensor $T^{\mu}_{\alpha\beta}$ [...] is obviously formally the simplest that our theory admits, the simplest characterization [of the] continuum will have to be linked to it, and not to the more complicated Riemann tensor.[43]

---

[41] Cf. Sauer (2006), p. 409.

[42] "[D]as Bestreben der Theoretiker [ist] darauf gerichtet, natürliche Verallgemeinerungen oder Ergänzungen der Riemannschen Geometrie aufzufinden, welche begriffsreicher sind als diese, in der Hoffnung, zu einem logischen Gebäude zu gelangen, das alle physikalischen Feldbegriffe unter einem einzigen Gesichtspunkte vereinigt" (Einstein, 1928a, p. 217). That the move to a teleparallel theory was inspired by the idea that a generalization of the axioms of geometry should produce the correct unified field theory is also illustrated by an unpublished manuscript in the Einstein archives; EA 2-111, "Über den gegenwärtigen Stand der Relativitäts-theorie", dated 1932, but presumably written earlier. Cartan brought his work to Einstein's attention in a letter, sent on 8 May 1929; their ensuing correspondence has been published in Debever (1979). Cartan subsequently took a strong interest in the unified field theory program. Their correspondence ended when Einstein moved away from the teleparallel theory.

[43] "Da dieser Tensor $T^{\mu}_{\alpha\beta}$ [...] offenbar der formal einfachste ist, welchen unsere Theorie zulässt, so wird an ihn die einfachste Charakterisierung [des] Kontinuums anzuknüpfen haben, nicht aber an den komplizierteren Riemannschen Krümmungstensor," (Einstein, 1928a, p. 221).

In his first publication on this geometry, however, no physical interpretation was given for the torsion. Nevertheless, the theory "should already be interesting due to the naturalness of the concepts introduced."[44]

This starting point of the teleparallel project can of course hardly be a surprise in light of Einstein's Oxford lecture, even if that lecture was still four years away. Similarly, it is easily captured in the diagram that he later drew for Maurice Solovine: we easily recognize the diagram's leap in Einstein's imaginative revision of the geometric axioms, a leap to a new set of axioms *A*, from which the deduction should begin. In a second article, Einstein thus derived ("in a completely simple and natural way"[45]) field equations for the gravitational field from a torsion tensor Lagrangian. In lowest order, these equations were the same as the lowest order approximation of the field equations of his 1915 theory. The theory also reproduced the vacuum Maxwell equations.

His colleagues, among them Wolfgang Pauli, nevertheless wanted to know what had become of the perihelion motion of Mercury and the gravitational bending of light; it was far from obvious how these were to be retrieved in the new theory.[46] Some results, for example those of Igor Tamm and Mikhail A. Leontowich, who were working with a revised version of the theory, seemed to indicate that the Schwarzschild solution could be recovered, promising perhaps the retainment of the motion of Mercury and gravitational deflection.[47] However, the deductions in the teleparallel theory led to problems that could not be resolved. A major set-back was that Einstein did not succeed in deducing non-singular static solutions that could be interpreted as electrons or light quanta. Another problem was that the theory did not appear to come with a uniquely determined set of field equations.[48]

Hermann Weyl returned to gauge symmetry eleven years after publishing his unified field theory. He did so in a paper that studied spinor theory in curved spacetimes. Now the gauge transformation was performed on the matter field ψ, not on the metric

$$\psi \to e^{ie\Lambda(x)}\psi, \qquad A_\mu \to A_\mu - \partial_\mu\Lambda(x). \tag{3.6}$$

The introduction of the local gauge symmetry in the phase factor of ψ enabled electromagnetism to be, in today's parlance, "derived" from the gauge principle: a Lagrangian density $L(\psi, \partial\psi)$ was made invariant under local transformations $e^{i\Lambda(x)}$ by changing all derivatives ∂_μ to covariant derivatives $\partial_\mu + iA_\mu$; Weyl

[44] "[...] Bestrebungen haben mich zu einer Theorie geführt, welche ohne jeden Versuch einer physikalischen Deutung mitgeteilt werden möge, weil sie schon wegen der Natürlichkeit der eingeführten Begriffe ein gewisses Interesse beanspruchen kann" (Einstein, 1928a, p. 217).

[45] "[...] ganz einfach und natürlich," (Einstein, 1928b, p. 224).

[46] W. Pauli to Einstein, 29 December 1929, see Pais (1982), p. 347.

[47] See the discussion in Goenner (2004), section 6.4.5.

[48] Sauer (2006), pp. 430–433; see also Vizgin (1994), pp. 254–256.

subsequently added a kinetic term $F_{\mu\nu}F^{\mu\nu}$ to the Lagrangian to retrieve Maxwell's equations with a current that was a function of ψ.[49]

Weyl believed that his former unified field theory did not compare favorably with his new use of the gauge principle:

This new principle of gauge invariance, that derives not from speculation but from experiment, seems to me [...] to show that electromagnetism is a necessary accompanying phenomenon, not of the gravitational field, but of the material wave field represented by ψ.[50]

The remark was not just a mild reconsideration of his earlier thoughts. It also contained veiled criticism of the many unified field theories that had been pursued since the publication of his own 1918 paper. By 1929, when Weyl wrote the above words, Einstein had become the most prominent proponent of this program,[51] so Weyl's criticism ought to be taken as directed at Einstein too. Pauli, who had not had a high opinion of Weyl's earlier theory – "Your earlier theory [...] was pure mathematics and unphysical; Einstein was justified in criticizing and scolding" – was now very much impressed by Weyl's work. He did not fail to remind Weyl that "the hour of your revenge has come."[52]

Weyl certainly did take revenge. He did so in a popular exposé of his recent work in *Die Naturwissenschaften*, a widely read and well respected journal.[53] One of the theories that he addressed in the article was his own 1918 theory, another was Einstein's 1928 teleparallel geometry. Proposals for Lagrangians in Weyl's original theory had only contained terms consisting of the gravitational Riemann tensor and the electromagnetic fields; similarly, Lagrangians for Einstein's teleparallel theory typically consisted of terms quadratic in the torsion.[54] Weyl now objected that neither of these theories had from their inception incorporated the matter field ψ, acting on the geometric structures.[55] Yet, recent diffraction experiments had established the wave character of matter beyond any doubt (these experiments, led

---

[49] For more on Weyl's procedure, see for example pp. 117–119 in O'Raifeartaigh (1997); Weyl's paper is (Weyl, 1929).

[50] "Es scheint mir [...] dieses nicht aus der Spekulation, sondern aus der Erfahrung stammende neue Prinzip der Eichinvarianz zwingend darauf hinzuweisen, dass das elektrische Feld ein notwendiges Begleitphänomen nicht des Gravitationsfeldes, sondern des materiellen, durch ψ dargestellten Wellenfeldes ist." (Weyl, 1929), on p. 246 in Chandrasekharan (1968).

[51] Unified field theory historian Hubert Goenner believes however that Einstein did not play the role of a conceptual innovator in the unified field theory program, despite having been its "central identification figure" because of his stature; Goenner's judgment is largely motivated by the fact that some of Einstein's proposals were anticipated by others (as in the case of Cartan); see Goenner (2004), section 11.

[52] "Als Sie früher die Theorie [...] machten, war dies reine Mathematik und unphysikalisch, Einstein könnte mit Recht kritisieren und schimpfen. Nun ist die Stunde der Rache für Sie gekommen." Pauli continued with: "[J]etzt hat Einstein den Bock des Fernparallelismus geschossen, der auch nur reine Mathematik ist und nichts mit Physik zu tun hat, und *Sie* können schimpfen!" Letter to Hermann Weyl, 26 August 1929, Doc. 235 in Hermann *et al.* (1979), on pp. 518–519.

[53] See Weyl (1931a). Weyl's essay was a written version of his Rouse Ball lecture, delivered at the University of Cambridge in May 1930.

[54] See Goenner (2004), section 4.1, for an overview of Lagrangians in Weyl's theory, and Sauer (2006) regarding proposals for Lagrangians in Einstein's theory.

[55] For commentary on Weyl's thought on matter, see Scholz (2006).

by Clinton J. Davisson in the United States and George P. Thomson in Britain,[56] were actually already some four years old) and Weyl believed that an additional wave equation for the ψ-field had become indispensable: "in the context of classical field physics, the state function ψ, the matter field, must be fitted in alongside gravity and electromagnetism. Not *two*, but *three* things must be reconciled."[57] According to Weyl, it was not the lack of atomistic solutions or even the non-probabilistic character that was the greatest shortcoming of the unified field theories, it was the absence of a matter field ψ.

Weyl did not show much clemency when expressing his dislike for Einstein's teleparallel theory. He believed its geometry to be, in fact, "unnatural"[58] and pointed out that it led to difficulties with the conservation laws. The achievements that his new gauge theory had conquered over the troubled unified field theories were numerous. For instance, the gauge factor $e^{i\Lambda(x)}$ was now complex valued, which met his earlier intuitions. Furthermore, the fields A_μ were expressed in the familiar electrodynamic quantity $e/2\pi h$ in the new theory, instead of in terms of some unknown cosmological quantity; Weyl argued that "to a healthy physical mind, that has not been spoiled by speculation, it will appear more sympathetic that the electrical field follows as the wake of the ship of matter and not that of gravitation."[59] However, he found it most important that his idea of gauge invariance had become incorporated into the Dirac theory of the electron – the theory that could account for the anomalous Zeeman effect and the fine structure of hydrogen. Following this rediscovery, he claimed that his "new" gauge principle had "grown from *experience* and sums up an enormous wealth of spectroscopical observations." Unified field theories, on the other hand, had become too alienated from observation and experience.

Weyl identified this weakness in his own earlier theory, and at least as much in Einstein's latest teleparallel contribution. He now believed that "all these geometrical capers [Luftsprünge] came too early." With the new form of the gauge principle, "we return to the firm ground of physical facts."[60] Einstein's Solovine schema-like leap to a teleparallel set of axioms – a *Luftsprung* that reflected the conviction that nature complies with the simplest mathematical formulation imaginable – had been a vain exercise, spoiled by speculation and lacking in solid contact with empirical

---

[56] See Davisson and Germer (1927); Thomson and Reid (1927); see also for example Jammer (1989), pp. 252–255.

[57] "[I]n den Rahmen der klassischen Feldphysik muss die Zustandsgrösse ψ, das Materiefeld, neben Gravitation und Elektromagnetismus eingefügt werden. Nicht *zwei*, sondern *drei* Dinge sind unter einen Hut zu bringen", (Weyl, 1931a, p. 51), emphasis as in original.

[58] "[U]nnatürlich," on p. 56 in Weyl (1931a).

[59] "Dem gesunden, von Spekulation nicht verdorbenen physikalischen Sinn ist wohl das auch viel sympathischer, dass das elektrische Feld dem Schiff der Materie und nicht der Gravitation als Kielwasser folgt." (Weyl 1931a, p. 58).

[60] "Das neue Prinzip ist aus der *Erfahrung* erwachsen und resümiert einen gewaltigen, aus der Spektroskopie entsprungenen Erfahrungsschatz," (Weyl, 1931a), p. 57, emphasis as in original; "Alle diese geometrischen Luftsprünge waren verfrüht, wir kehren zurück auf den festen Boden der physikalischen Tatsachen," (Ibid., p. 56).

facts. Weyl by now disapproved of the classical geometrization and unification program, but if further pursued, it at least had to start with "the geometrization of the matter field; if one succeeds at that, the electromagnetic field will be automatically included as an encore to the enterprise."[61]

Eventually Einstein moved away from the teleparallel theory. He first returned to Kaluza–Klein theory, a theory that he had already studied before, this time producing two papers, co-authored by his mathematical assistant Walther Mayer.[62] Weyl responded to this work in a letter to Einstein:

I did not care for your teleparallelism at all. Your new *Ansatz* looks much more promising to me and appeals much more to my mathematical heart. [...] Perhaps it will tempt me to also think again about the universal world geometry. But it seems absolutely necessary to me to include the Schrödinger–Dirac ψ. And "gauge invariance," as I currently understand it to connect ψ (and not ds) with the electromagnetic potential, very well indicates in what direction this should take place.[63]

When Einstein's first paper on "projective Kaluza–Klein theory" appeared, he himself qualified it as lacking promise, since it appeared to "founder on the problem of matter and quanta."[64] The second paper constructed a Lagrangian which consisted not only of the Einstein–Hilbert and Maxwell terms, but additionally had a matter term. Einstein and Mayer remarked that they had not yet been able to address the wider applicability of this theory, and a follow-up article never appeared.

The next publications by Einstein and Mayer were papers on spinors, "semivectors" and an unquantized version of the Dirac theory; these will be addressed in Chapter 5. They started out immediately from the "Schrödinger–Dirac" matter field.[65] Apparently Einstein now also felt that one could no longer neglect its independent existence, if not for empirical reasons, then at least because of the enhanced theoretical understanding of the gauge principle due to the work of Weyl.

Einstein's new perspective on matter wave fields can be traced back to at least as early as the Solvay conference of 1930. He wrote to Walther Mayer from Brussels that:

[61] "Auf die Geometrisierung des Materiefeldes also müsste man ausgehen; wenn man mit ihm reüssiert, geht das elektromagnetische Feld von selber als Zugabe in den Handel ein." As in Weyl (1931a), p. 58.

[62] See Einstein and Mayer (1931, 1932a).

[63] "Ihren Fernparallelismus habe ich gar nicht gern gehabt. Ihr neuer Ansatz scheint mir viel aussichtsreicher und ist meinem mathematischen Herzen viel sympatischer. [...] Vielleicht reizt mich das dazu, auch einmal wieder über die universelle Weltgeometrie nachzudenken. Mir scheint es doch absolut nötig, das Schrödinger-Dirac'sche ψ mit hineinzunehmen. Und die "Eichinvarianz", wie ich sie jetzt auffasse, als etwas, das ψ (und nicht ds) mit den elektromagnetischen Potentialen verbindet, deutet ja wohl an, in welcher Richtung das zu geschehen hat." 22 June 1932, Hermann Weyl to Einstein, EA 24-098.

[64] "Die Konstruktion scheint am Problem der Materie und dem der Quanten zu scheitern." From "Der Gegenwärtige Stand der Relativitätstheorie," lecture at the University of Vienna, October 14, 1931, published as Einstein (1932), on p. 442. In his correspondence with Ehrenfest a month before his lecture, Einstein still said that he expected much from the theory; see Einstein to P. Ehrenfest, 17 September 1931, EA 10-221. For more on the projective five dimensional approach, see Goenner (2004), section 6.3.2.

[65] Before moving to the Dirac theory, Einstein and Mayer discussed generalizations of Schrödinger-type equations to curved spacetimes, but these ideas never found their way to a publication. See EA 18-129 through EA 18-136, all from August 1932.

The essence of quantum mechanics lies in the existence of the de Broglie waves, if a field-like interpretation is at all possible. [...] It certainly has great appeal to try to attribute a physical reality to the de Broglie waves, in particular as these waves document themselves so directly in crystal interference experiments.[66]

When Einstein elaborated on his ideas in the same letter, the "de Broglie waves" were still wave solutions of the fields of the teleparallel theory, and he hoped to retrieve the Schrödinger equation in an approximation from the teleparallel field equations. But in their work on semivectors, Einstein and Mayer would indeed start out with the Dirac field as an independent entity in the field equations.

In this section we have seen Weyl criticize Einstein – and from a Whiggish point of view, rightfully so: the work of Weyl from the 1920s is regularly honored as a constitutive contribution to modern gauge theory,[67] whereas Einstein's teleparallel theory has been largely relegated to the peripheries of theoretical physics. Einstein at this time also greatly appreciated Weyl's work, and we will see him trying to understand Weyl's textbook on the mathematics of quantum theory when he was struggling with his semivectors. To Weyl, Einstein of course always remained a source of inspiration – Weyl, who would also become a refugee of National-Socialist Germany, chose to go to Princeton's Institute for Advanced Study partly because that was where Einstein had decided to go:

For me it is very important, in particular at the start, who are the men that will collaborate in the institute. Flexner[68] has told me (of course I will not tell anyone!) that he has invited you and has the prospect of winning you over. He has also authorized me to correspond confidentially with you on this issue. May I ask you how you feel about the matter? If namely you were to go there, then my decision to do the same is as good as made![69]

Yet, although Einstein and Weyl were colleagues for nearly twenty years in Princeton, a collaboration did not ensue. Weyl's work in mathematical physics continued

[66] "Die Essenz der Quantenmechanik liegt in der Existenz der de Broglie-Wellen, wenn überhaupt eine feldartige Interpretation möglich ist. [...] Es hat sicher einen grossen Reiz zu versuchen, den de Broglie-Wellen physikalische Realität zuzuschreiben, zumal sich diese Wellen bei den Kristall-Interferenz Verscuchen so unmittelbar dokumentieren." Einstein to Walther Mayer, 1 November 1930, EA 18-077; Einstein likely is referring here to the results of Davisson and Germer (1927); their electron diffraction pattern of a nickel crystal confirmed the ondulatory character of matter conjectured by de Broglie. For similar comments, see also Einstein to Walther Mayer (20/22 October 1930, EA 18-075): "The conference is very demanding, but it is clear that the quantum mechanical theory represents such exact characteristics of reality that it must contain a large portion of truth. So it is in any case right to study Dirac. Maybe this study will lead us to a possible *Ansatz*." ("Der Kongress ist sehr anstrengend, aber es zeigt sich, dass die quantenmechanische Theorie so feine Züge der Realität darstellt, dass sie ein grosses Stück Wahrheit enthalten muss. Es ist also jedenfalls richtig, wenn wir den Dirak [sic] studieren. Vielleicht führt uns dies Studium auf einen möglichen Ansatz.")
[67] According to field theorist Lochlainn O'Raifeartaigh's history of gauge theory (1997), p. 107, the 1929 paper by Weyl is "one of the seminal papers of the [twentieth] century."
[68] Abraham Flexner (1866–1959) was the first director of the Institute for Advanced Study.
[69] "[Es kommt] für mich, gerade für den Anfang, sehr darauf an, wer die Männer sind, die im Institut zusammen-wirken werden. Flexner hat mir erzählt (natürlich sage ich es niemanden weiter!), dass er auch Sie aufgefordert habe und Aussicht bestehe, Sie zu gewinnen. Er hat mich auch ermächtigt, mit Ihnen vertraulich darüber zu korrespondieren. Darf ich Sie fragen, wie Sie zu der Angelegenheit stehen? Wenn nämlich Sie hingehen, dann ist die Sache auch für mich im gleichen Sinne so gut wie entscheiden!" Weyl to Einstein, 22 June 1932, EA 24-097.

to stay focused on the mathematics of the quantum theory, whereas Einstein would only touch tangentially upon that subject during the semivector episode.

When we follow Einstein's advice and look at his deeds, we readily recognize the words he would soon speak in Oxford: unified field theories were permeated by the search for mathematical naturalness and logical simplicity. Einstein's epistemology implied that one could come to true understanding and to true advancement of theoretical physics foremost by following these research maxims. When Weyl criticized Einstein's program, he objected that the unified field theory methodology had produced theories that lacked any contact with the real world of experience: all that had been achieved were just vain "*Luftsprünge.*" Particularly telling for the inauspiciousness of Einstein's approach was that its theories had ignored the matter field and the wave equation, the accuracy of which had meanwhile been firmly established by a slew of experiments.

4

Experiment and experience

Even though Einstein had proclaimed in Oxford that "all knowledge of reality starts from experience and ends in it," we have seen that philosophically, more or less, he gradually moved from empiricism to realism, while his research became dominated by the unified field theory program. As we just saw, the program was criticized as it seemed to recede from the world of experience. This aspect, however, was identified by Einstein as a general trait of the search for unification. One may recognize it in the diagram that we drew in the preceding chapter, Figure 3.1: the distance between the concepts employed at the most fundamental level (A') and direct experience (E) grows. The unified theory "pays for its higher logical unity by having elementary concepts [...], which are no longer directly connected with complexes of sense experiences." Einstein found this evolution perhaps to be regrettable, but nonetheless something that one can only resign oneself to.[1]

A superficial glance at Einstein's professional career seems to reinforce further the image of the scholar who withdrew ever more into the ivory tower of mathematical principles: Einstein started off examining technological patents in the Bern patent office, yet ended as the sage who intuited unification axioms in the School of Mathematics of Princeton's Institute for Advanced Study. In this chapter, we wish to address the broader development in Einstein's relation to experiment and experience; we hope that this may give a complementary perspective to his turn to unified field theories, and thus contribute to a fuller understanding of the evolution in his thinking. We will first give a brief overview of his attitude to, and involvement with, actual experimentation.

[1] As in Einstein (1936a), on p. 323 in Einstein (1994).

4.1 Einstein and experimentation

When Einstein was a young man, his interests focused foremost on the empirical side of physics. According to his autobiography he had the opportunity to study with "excellent" teachers in mathematics at the ETH in Zurich (Einstein mentioned in particular Adolf Hurwitz and Hermann Minkoswki), but he "worked most of the time in the physical laboratory, fascinated by the direct contact with experience"; his interest in "knowledge of nature" was "unqualifiedly stronger" than in mathematics.[2] Einstein had furthermore grown up around his family's electrotechnical businesses, and his secondary school of choice in Aarau, Switzerland, was known for its fine physical cabinet, which even underwent an expansion while he was a student there.[3]

Einstein's interest in the empirical persisted after his graduation from the ETH, even though he had developed into a theoretically inclined physicist. His 1905 papers were written while he was employed as patent officer, and an emphasis on the empirical was not only in evidence in the semblance between Einstein's events and Mach's elements of sensation in the article on relativity. The light quantum paper and Einstein's treatment of Brownian motion connected in a direct way to experimentation: the first for instance connected through its discussion of the photo-electric effect, and the second included an experimental procedure for the determination of Avogadro's number. Conversely, the mathematics of these articles was fairly light.[4]

Einstein's most engaged involvement with engineering and experiment may have been his design of a "little machine" (his "*Maschinchen*") that could multiply and thus render measurable tiny potential differences.[5] He could show that such tiny potential differences ought to occur in a capacitor due to molecular fluctuations,[6] and as his analysis was closely related to his account of Brownian motion, the *Maschinchen* promised to deliver a more or less direct confirmation of his theory for the latter effect.

Conrad Habicht, a secondary school teacher and personal friend of Einstein (Habicht had also been a member of the Academia Olympia), started the construction of the *Maschinchen* in the summer of 1907. He was helped by his brother Paul, who was a professional machine technician. While they were trying to build Einstein's apparatus, the latter could barely contain his "*murderous* curiosity," and

---

[2] As in Einstein (1949a), pp. 14–15.
[3] See Fölsing (1998), p. 37.
[4] Cf. Norton (2000), p. 141. For the discussion of the photo-electric effect, see Einstein (1905a), pp. 145–147; for the determination of Avogadro's number, see Einstein (1905b), pp. 559–560.
[5] On Einstein's *Maschinchen*, see a fine article by Ad Maas, curator at Museum Boerhaave in Leiden (Maas, 2007). Maas recently uncovered a copy of the *Maschinchen* in the Boerhaave depots.
[6] See Einstein (1907a).

anxiously awaited news that would inform him of any progress.[7] Initially the Habichts were unsuccessful, however, and Einstein impatiently sought to engage others in the project. As a result, he would have three copies of the instrument in hand by the end of 1908, one of which had indeed been constructed by Paul Habicht.

By then Einstein had already rolled up his sleeves: he had started testing and experimenting in the summer of 1908, later joined again by the Habicht brothers. His Bern apartment came to contain his own "small laboratory for electrotechnical experiments."[8] In the end, however, the *Maschinchen* was neither an entrepreneurial nor a scientific success. It suffered from operational difficulties and the competition soon produced instruments with better sensitivity. Einstein's theory of Brownian motion would first be confirmed by experiments conducted by Jean Perrin in 1908.

Another instance in which Einstein got his hands dirty experimenting was his collaboration with Dutchman Wander J. de Haas, with whom he worked during a lull in his work on general relativity in the winter of 1915. They intended to establish the existence of "Ampère's molecular currents," that is, the idea that the field of a ferromagnet or paramagnet is the result of currents on a molecular scale. Their experiments were conducted in the laboratories of the *Physikalisch-Technische Reichsanstalt* in Charlottenburg, near Berlin; de Haas was an experimentalist with a recent doctorate from Leiden University (he was actually Lorentz's son-in-law) and had joined the *Reichsanstalt*'s staff for a short period.[9]

Einstein and de Haas wanted to test Ampère's hypothesis by making a metal bar rotate by magnetizing it (see Figure 4.1). The underlying idea was simple. Ampère's molecular currents were attributed to electrons that circled an atomic nucleus, and the magnetic moments of these tiny currents would also be aligned by magnetizing the rod. This implied that the angular momenta of the electrons would also be aligned. Since the overall angular momentum had to be conserved, the bar would then start to rotate in a direction opposite to the directed trajectories of the electrons.

The observation of such a rotation would qualitatively confirm Ampère's hypothesis. But Einstein and de Haas wanted more: they also wanted to establish the gyromagnetic ratio of the motion of an electron quantitatively. This ratio is given by

$$\frac{L}{M} = \frac{2m}{e} \times \frac{1}{g}.$$

[7] "*Mörderische* Neugier"; Einstein to Paul and Conrad Habicht, 2 September 1907, Doc. 56, p. 72 in Klein *et al.* (1993), emphasis in original.

[8] Einstein to Johannes Stark, 14 December 1907, Doc. 132, p. 152 in Klein *et al.* (1993); see also Einstein to Albert Gockel, 3 December 1907, pp. 150–151 in Klein *et al.* (1993).

[9] The following discussion is largely based on the treatment of the Einstein–de Haas collaboration in Galison (1987), chapter 2; see also the editorial note "Einstein on Ampère's molecular currents" in Kox *et al.* (1996), pp. 145–149. References to original publications are Einstein and de Haas (1915) and Einstein and de Haas (1915–1916).

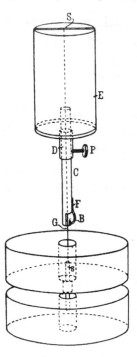

Figure 4.1. In Einstein and de Haas's experimental arrangement a small metal rod S, depicted in the bottom of the diagram, hung between two coils. The coils produced an alternating magnetic field, which led to an oscillating rotation of the rod. Two small mirrors, on which a light-beam fell, were connected to the rod. The amplitude of the oscillation was determined by reading off the position of the reflected light-beam from a screen. This amplitude was a measure for the couple that the alternating magnetization caused in the rod; from this couple, in turn, the gyromagnetic ratio of the motion of the electrons followed. From Einstein and de Haas (1915), p. 160.

L and M are, respectively, the angular momentum and magnetic moment of the electron, m and e are its mass and charge. "g" is known as the "g-factor," and in Einstein and de Haas's theoretical model[10] $g = 1$. This value was based on the assumption that the electron was a charged point particle circling an atomic nucleus; it had no further properties.

Einstein and de Haas's first measurements gave results that were equivalent to $g = 1.4$. It was soon clear that a systematic error had crept in: the magnetic field used had been too weak. A second, improved measurement gave a more satisfactory

---

[10] Einstein and De Haas never explicitly included the g-factor in their calculations, but their relations amounted to the same thing; we follow Peter Galison's treatment in introducing g in the discussion of the Einstein–de Haas experiment; see Galison (1987), chapter 2.

result: $g = 1.02 \pm 0.10$. Einstein and de Haas concluded that Ampère's hypothesis had been "fairly confirmed by our observations."[11]

During the 1920s, however, an increasing number of results by other experimenters appeared that suggested that the true value of g ought to be closer to 2. The discovery of electron-spin, in 1925, could also give a theoretical explanation of that value: the rotation of the rod was due to the alignment of the spins of its electrons rather than their orbital angular momentum.

Why did Einstein and de Haas measure a value for g that was half the value established later? Their experimental arrangement in all likelihood still allowed for too many systematic errors. The earth's magnetic field, for instance, was a disturbing factor. Einstein and de Haas had tried to eliminate it, but the order of magnitude of its possible disturbances was disquietingly big. They further had not measured the magnitude of the magnetic field that they had applied, but had calculated it on the basis of the constants of the coils. Historian Peter Galison has persuasively argued that when Einstein and de Haas found $g = 1.02 \pm 0.10$, this was such a good confirmation of their prior expectations that they stopped looking for any possible further systematic errors. Experimentation simply ended when the theory had been confirmed.[12]

What kind of an image of Einstein is suggested by the above encounters with experimentation? An Einstein that threw himself into experimental projects and filled his apartment with charge multipliers of his own design, or was busily engaged with coils and rods in the rooms of the *Reichsanstalt*. Theoretical bias seems to have obstructed a result that would still hold today in the case of the Einstein–de Haas experiment – yet, Einstein had translated a simple theoretical idea into a straightforward experiment and was diligently trying to ferret out an experimental verdict on Ampère's molecular currents. These episodes in experimental tinkering of the younger Einstein suggest a different approach to physics than the austere search for unifying mathematical principles that typified his later physics.

Yet, this is of course not a fair comparison. Firstly, Einstein primarily published theory, both in his early and in his later career. It is furthermore – reflective of his position on the realism versus empiricism dichotomy – impossible to make clear black and white distinctions between a young and an old Einstein in his relation to actual experiment. His engagement with technology and experiment would never be entirely absent: in 1923, for instance, he published a (minor) experimental paper on the permeability of filters, and he kept involved in engineering projects well into his later Berlin years. Even in Princeton he would still engage with the practical, as may be exemplified by his 1935 patent application for an automated

[11] As in Einstein and de Haas (1915–1916), p. 711.
[12] See Galison (1987), in particular pp. 72–74.

exposure-regulation scheme in photographic cameras, filed together with fellow amateur inventor Gustav Bucky.[13] Conversely, the young Einstein had at times valued theory more than experiment. In 1907, for example, he had not been convinced by Walter Kaufmann's data that seemed to confirm formulations of electrodynamics by Max Abraham and Alfred Bucherer rather than his own. In this instance he exhibited such a strong confidence in theory that one is reminded of the older Einstein.[14]

A gradual receding from experiment and observation does nevertheless appear to have been a characteristic element of the development of Einstein's practice in physics – one observes a gradual decoupling of the pragmatic and empirical side of the field of physics from the core of his professional science: his efforts in theory construction. As we saw already in Chapter 3, this is evident if we look at his unified field theory efforts: Weyl complained about a lack of interaction with, or even consideration of, experimental findings when Einstein was formulating unification attempts in Berlin. Such a complaint could not easily have been made against his earlier theorizing. His field theory constructions in Princeton continued to approach a complete detachment from contemporary experimentation: Einstein had, in the unflattering words of IAS director J. Robert Oppenheimer, in his quest "to realize the unity of knowledge [...], turned his back on experiments."[15]

In the 1920s Einstein's principal research did on occasion still connect to experimentation. One of the primary motivations for appointing Einstein to a special professorship at Leiden University in 1920 was a hoped for involvement with its cryogenic laboratory, headed by Heike Kamerlingh Onnes. Indeed, Einstein made specific suggestions for a theory of superconductivity, and some of its implications were promptly tested by Leiden experimentalists.[16] But Einstein's engagement with the laboratory quickly faded during the 1920s, just as his visits to Leiden became increasingly less frequent.

Over the years, his personal involvement with the actual execution of experiments declined – so much seems clear if one looks at the few instances that have been documented in the published record of Einstein's oeuvre.[17] This disengagement may be seen on comparing his role in four experiments of his own design: the

---

[13] The 1923 paper is Einstein and Mühsam (1923); for Einstein's involvement in Berlin with a gyrocompass program and a silent refrigerator project see Fölsing (1998), pp. 596–599; see Isaacson (2007), p. 435 for his collaboration with Bucky. In 1922, Einstein briefly considered returning to an occupation in technology, but this was mostly motivated by the strain that his very public presence produced that year; see Einstein's letter to Hermann Anschütz-Kaempfe, 16 July 1922, EA 80 720.

[14] For Einstein's reaction to Kaufmann's work, see for example Hentschel (1992).

[15] J. Robert Oppenheimer, as in an interview in Paris with *L'Express*, 1965; see Schweber (2008), p. 282.

[16] On Einstein's work on superconductivity and the tests proposed by him, see Einstein (1922) and Sauer (2007a).

[17] It is of course possible that more instances of Einstein interacting with experimental physics will surface in the coming years, in particular as the Einstein Papers Project assesses more unpublished source material. However, we do not expect this to alter the overall development discussed here.

Einstein–de Haas experiment of 1915 that we just discussed, an experiment on the wave and particle aspects of light from 1921, and another two experiments on the same subject from 1926 (the latter were the "last in this genre," biographer Abraham Pais found[18]). In the case of the Einstein–de Haas project, as we saw, Einstein joined Wander de Haas at the *Reichsanstalt* in the nearby suburb of Charlottenburg. In 1921, in the same laboratory, experimentalists Hans Geiger and Walther Bothe conducted an experiment on the light quantum that had indeed also been designed by Einstein, but this time he did not join the experimentation, even though it was carried out at the same nearby location.

The experiment was unsuccessful, principally because Einstein's original theoretical analysis had actually been flawed.[19] Could the fact that Einstein did not himself participate in its execution count as an early indication of his gradual withdrawal from experimentation? It very well might, but other reasons could of course explain his non-involvement; for instance, his schedule may simply have become too busy. He also attributed the experiment great value, so any philosophical detachment from experimentation would still not have been quite complete.

The two experiments on the nature of light thought out in 1926 were to be conducted by Emil Rupp (see Figure 4.2) of the University of Heidelberg, and again Einstein stayed away from the actual experimentation. This time there were obvious reasons for him to do so, one of course being the large distance separating Berlin from Heidelberg. Most important, however, was the fact that Rupp's institute was headed by Philipp Lenard, a rabid anti-Semite and vociferous opponent of relativity.[20]

There was however yet another, rather practical reason for Einstein to stay away from Heidelberg in 1926, and this factor helps to explain his gradual receding from the laboratory during the larger period. Einstein said that he felt increasingly incompetent to judge the internal criteria adopted by experimentalists. As he would say in 1927: "being a theorist, I may not barge into the craft of the gentlemen experimenters."[21] Einstein had apparently become uncomfortable in the laboratory, as experimentation required ever greater specialist skills. His sentiment was of course a reflection of the general development of specialization in physics, and in particular of the forking of the profession into theoretical and experimental subdisciplines that was taking place in these years. However, that development

---

[18] See Pais (1982), p. 329.

[19] The experiment is outlined in Einstein (1921b); for a discussion, see Klein (1970), pp. 8–13, and Tauschinsky and van Dongen (2008).

[20] See Einstein to Emil Rupp, 18 April 1926, EA 70 703. On the relation between Einstein and Lenard, see for example Schönbeck (2000), van Dongen (2007c).

[21] "[I]ch Theoretiker darf den Herren Experimentatoren nicht ins Handwerk pfuschen." Einstein in an interview with Richard Wolf, *Deutsche Allgemeine Zeitung*, 27 November 1927, reprinted in Hentschel (1992), on p. 610; see also Hentschel's discussion on pp. 612–613.

Figure 4.2. Emil Rupp (by permission of the Niedersächsische Staats- und Universitätsbibliothek Göttingen).

seems to exhibit itself in Einstein's case in a particularly pronounced fashion, given the above words.

The Einstein–Rupp experiments led to a rather extreme repercussion of Einstein's gradual distancing from the practice of experiment. Rupp's results have been established to have been the product of nothing less than fraud. Einstein, however, had earlier presented them to the Prussian Academy as the result of a joint project, and saw to their publication in the Academy's proceedings; Rupp's paper immediately followed a theoretical article by Einstein that outlined the experiments, and reprints circulated that combined both papers in a single booklet.[22]

The Einstein–Rupp experiments were, as indicated, originally designed to resolve the apparent conflict regarding the wave and particle nature of light. Einstein initially believed that the experiments were going to show that light emission occurred instantaneously, as one might expect for particle-like light quanta. The experiments would do this by disproving wave theory predictions for their outcome – these hinged on the presumption that light emission is a process that is extended in time. Both experiments used a Michelson interferometer and a canal ray light source – excited atoms that move with a more or less homogeneous velocity together in a beam and emit light. In the first experiment, the light from the canal ray particles first passed through a wire grid before it entered the interferometer ("wire grid experiment," see Figures 4.3 and 4.4); according to the wave theory, varying the path difference in the interferometer would show a variability in the visibility of

---

[22] The following is based on the more extensive discussion of the Einstein–Rupp experiments contained in van Dongen (2007a,b); for more on Emil Rupp's fraudulent career, see also French (1999). The joint Einstein–Rupp articles referred to are Einstein (1926b), Rupp (1926b).

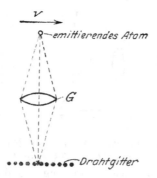

Figure 4.3. The wire grid experiment of 1926: an atom, traveling in a canal ray beam with velocity v parallel to a grating G, emits light onto the grating. After passing the grating, the light falls on a Michelson interferometer. Taken from Einstein (1926a), p. 300 (by permission of Springer Science+Business Media).

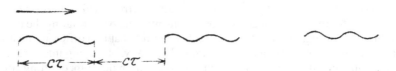

Figure 4.4. Einstein's wire grid experiment: in a wave picture, the light emitted is cut up as the emitting atom travels along the grating. The Michelson interferometer then produces a varying visibility of the interference pattern for varying path differences: as the wavepackets are $c\tau$ long (τ being the time it takes the atom to pass an opening in the grating and c the velocity of light), the interference fringes should be clearly visible for path differences equal to $2n \times c\tau$, whereas they should be absent for values of $(2n+1) \times c\tau$, with n an integer. Initially, Einstein expected that the quantum picture of light implied that this variation in the visibility of the interferences would not exhibit itself. He thought that the emission of a quantum would be an instantaneous process, and that somehow (Einstein did not explain), if the emission of a light quantum were instantaneous, the wave associated with a light quantum would not be cut up. He soon however changed his mind, expecting the variation to be observed in any case; the experiment would then in the end not be able to decide between quantum and classical emission scenarios. From Einstein (1926a), p. 300 (by permission of Springer Science+Business Media).

the interference pattern. Einstein initially expected that this variability would not be observed owing to the quantum nature of light,[23] but he had changed his mind by the time Rupp started conducting the experiment.

In the other experiment, in which the canal ray beam was put in the focal plane of the lens and the wire grid removed (see Figure 4.5), the rotation of one of the interferometer's mirrors over a particular angle was to offset disturbing consequences

---

[23] See Einstein (1926a).

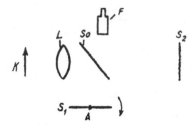

Figure 4.5. The rotated mirror experiment, from Einstein's publication in the Academy proceedings (Einstein, 1926b, p. 337). In the figure, K is the canal ray light source, consisting of excited atoms moving up and emitting light to the right. L is a lens and S_0, S_1 and S_2 make up the Michelson interferometer's beam splitter and mirrors respectively. The interference pattern is observed with telescope F. Due to the motion of the atoms in the canal ray beam, blue- and redshifting occurs in the emitted light; light arriving at the top of lens L is slighty shifted to the blue, light arriving at the bottom is slightly redshifted. This disturbs the interference pattern that should form in the interferometer, as the path difference per wavelength varies in directions parallel to the interferometer's mirrors. This variation can be compensated for by a slight rotation of one of the mirrors. After rotating the mirror, the light that is brought to interference has been emitted at two different locations and at different times in the canal ray beam K. Naturally, observing interference is then taken to imply that the light has been emitted by a single atom, emitting coherent light as it moves up in the canal ray beam, i.e., it implies that light emission is a process that is extended in time. In the diagram reproduced above, Einstein, surprisingly, misdrew the angle of rotation; according to his theory he should have drawn the direction of rotation in the opposite direction.

of the Doppler effect; the latter were due to the beam velocity of the canal ray atoms and their disturbance was brought to bear by the new position of the lens relative to the beam. This experiment was known in the literature as the "rotated mirror experiment," and Einstein again expected the classical theory to be confirmed, i.e. to see interference restored by the rotation through a particular angle of one of the interferometer's mirrors. Initially, as said, he had expected an outcome in line with a quantum-like instantaneous emission, which would have implied that no interference upon rotation of the mirrors would be observed. Nevertheless, as in the case of the wire grid experiment, he had changed his mind before Rupp actually got to work.

Rupp indeed soon claimed to have confirmed Einstein's expectations for both experiments. However, knowledgable experimentalists had already doubted the suitability of Rupp's experimental arrangement even before his results on Einstein's experiments were out,[24] and these too were soon found to be deeply flawed.

In the case of the wire grid experiment, Rupp claimed to have seen interferences at path differences of tens of centimeters. The canal ray light sources that were

---

[24] See Atkinson (1926) and Rüchardt (1926).

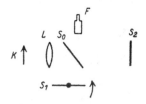

Figure 4.6. In 1935 experimentalists Walther Gerlach and Eduard Rüchardt published this corrected version of Einstein's diagram in an article that argued that Rupp had falsely claimed to have carried out the rotated mirror experiment. From Gerlach and Rüchardt (1935), p. 124.

available at the time, however, had such an inhomogeneous velocity distribution – and hence an additional, wide spread in wavelengths of the emitted light – that they could really only allow interference to be observed for path differences of less than a millimeter. This was much too short to be in a position to observe the variability that Einstein had predicted. The beams were also too inhomogeneous to allow for a clear angle of rotation to be measured to confirm Einstein's analysis of the rotated mirror experiment, contrary, however, to Rupp's results (see Figure 4.8 below). In fact, as was established a decade later, Rupp's published angle of rotation for the mirror even pointed in the wrong, opposite direction (see Figures 4.5, 4.6 and 4.7). All in all, Rupp's contemporaries were convinced that he had forged his results, and historical research has strongly confirmed these suspicions.[25]

Why did Einstein go along with Rupp's experimental work? Einstein of course believed that Rupp was reporting his results truthfully. He was no practitioner of canal ray experiments, which made judgment difficult, and we have already seen him admit that he had become reluctant to tread on the experimentalists's turf. Rupp further usually communicated his findings with a fairly professional presentation that provided considerable, hence convincing, circumstantial detail. On the other hand, his results often did not make sense and he repeatedly changed his data and the clarification for his data after persistent questions from Einstein, and criticism from others. This however did not lead Einstein to doubt Rupp's work.

Another circumstance to be factored in is that Einstein was convinced of the correctness of his own theoretical analysis. He would therefore have expected Rupp only to find complete agreement with his predictions: their correspondence shows that Einstein corrected Rupp long enough until he got the results that he expected. Once he believed that Rupp had found such confirmation, he apparently felt no further need to scrutinize the latter's work – or, for that matter, to attend to publications that questioned Rupp's data and experimental arrangement

[25] On path differences, see Atkinson (1929), Straub (1930), for the argument that Rupp had rotated his mirror in the opposite direction, see Gerlach and Rüchardt (1935); for historical analysis see van Dongen (2007a,b).

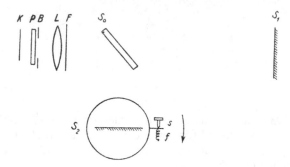

Figure 4.7. Emil Rupp's diagram for the rotated mirror experiment, from Rupp (1926b), p. 347. The canal ray beam at K was moving up. With the mirrors in the above relative positions, Rupp should have rotated the mirror in the opposite direction. However, his angle of rotation confirmed Einstein's incorrectly drawn angle from the latter's accompanying publication. (In this diagram F is a filter that should limit the light to the desired wavelength range, B is a stop and P is a glass plate.)

(in particular a highly critical and very visible publication by British spectroscopist and astrophysicist Robert Atkinson deserves mention[26]). Thus theoretical bias, as in the case of the Einstein–de Haas experiment, had likely further eroded a critical attitude in Einstein's assessment of Rupp's work.

Finally, an increasingly less than critical attitude to the practice of experiment is also suggested by Einstein's correspondence on the nature of light. In September of 1925, six months before Einstein's first contacts with Rupp, Paul Ehrenfest wrote to Einstein expressing his hope that an upcoming conference in Leiden would be the occasion for discussions between the latter and Niels Bohr; Ehrenfest wished that these discussions would probe the problems of the quantum, "particularly [...] regarding experiments that you always think out on the frontier between 'waves and particles'."[27] Einstein's reply is revealing:

I no longer think about experiments on the boundary between waves and particles; I believe that this was a vain effort. Inductive means will never get you to a sensible theory, even though I do believe truly foundational experiments, like Stern and Gerlach's and Geiger and Bothe's, can be a real help.[28]

---

[26] It had come out in the quite visible journal *Naturwissenschaften* (Atkinson, 1926). This article appeared a little before Einstein submitted his work with Rupp to the Prussian Academy and questioned results that Rupp had obtained earlier with the same experimental setup (Rupp, 1926a); for discussion, see van Dongen (2007a), pp. 76–80, 85–86.

[27] "Ganz besonders interessant ist mir, was die Discussion zwischen Dir und Bohr betreffs der Experimente liefern wird, die Du immer an der Grenze zwischen „Wellen und Corpuskeln" planst." Paul Ehrenfest to Einstein, 16 September 1925, EA 10 110.

[28] "An Experimente an der Grenze Wellen-Korpuskeln denke ich nicht mehr; ich glaube, dass dies eine verfehlte Bemühung war. Auf induktivem Wege wird man wohl nie zu einer vernünftigen Theorie kommen, wenn ich auch glaube, dass ganz prinzipielle Experimente wie das vom Stern-Gerlach oder Geiger-Bothe ernsthaft nützen können." Einstein to Paul Ehrenfest, 18 September 1926, EA 10 111.

The Einstein–Rupp experiments were of course intended to probe exactly the wave-particle frontier, and they can further be seen as prime examples of the kind of "foundational experiments" that Einstein still appreciated (in the above letter he was most likely referring to Geiger and Bothe's 1924 test of the BKS theory, and not their failed 1921 collaboration). The above comment, however, also suggests a certain limited regard for the epistemological merit of experimentation for the creative theorist. Einstein seemed to imply that valuable experiments are those that yield a clear "yes" or "no" answer to a theoretical question of principle. He did not expect much from inductive searching in the realm of experience.

A decade later, Einstein, now in Princeton, exchanged letters on the Einstein–Rupp experiments with a former close colleague from Berlin, Max von Laue. Rupp's career had come to a spectacular collapse in 1935 after continued and increasingly intense charges of fraud; these had come to challenge a whole slew of his latest experiments (his canal ray experiments had already come under heavy attack in 1930; see Figure 4.8). He issued a public withdrawal of five of his most recent publications, which was promptly followed by an ordinance of the German Physical Society to its journals no longer to accept any of Rupp's papers.[29] There were renewed discussions about the Einstein–Rupp experiments in the ensuing fallout. Von Laue had come to believe not only that Rupp had forged his data, but now also that Einstein's original theory for the experiments had been wrong. Einstein defended his original arguments against von Laue's objections without even once mentioning the name of Emil Rupp:

I do not consider my considerations of those days to be superfluous or false. I even believe that they might still be fairly interesting. Because in my opinion, we today still lack a theory that can be taken seriously. Pardon me for elaborating, but I see that you have not appreciated the point that makes my considerations of those days meaningful. Of course, also back then they did not require any confirmation by experiment.[30]

To Einstein in 1936 his original theoretical analysis had been obviously correct, so no experiment had ever been necessary. Theoretical bias, the professional reservations of the theorist vis-à-vis the expertise of the experimenter, and an increased philosophical indifference had earlier prevented him from scrutinizing Rupp. But now the purported self-evidence of the experiments allowed him – likely in part in embarrassment – to repress them altogether; Einstein indeed emphasized to von

---

[29] See van Dongen (2007a), p. 111.
[30] "Ich halte daher meine damaligen Überlegungen nicht für überflüssig oder falsch, ja ich glaube sogar, dass sie immer noch ein bescheidenes Interesse beanspruchen dürften. Denn eine ernst zu nehmende Theorie besitzen wir meines Erachtens auch heute noch nicht. Verzeihe die Ausführlichkeit. Aber ich sehe, dass Du den Standpunkt nicht erfasst hattest, von dem aus meine damaligen Überlegungen sinnvoll sind. Natürlich bedurften sie auch damals keiner Bestätigung durch den Versuch." Einstein to Max von Laue, 29 August 1936, EA 16 113.

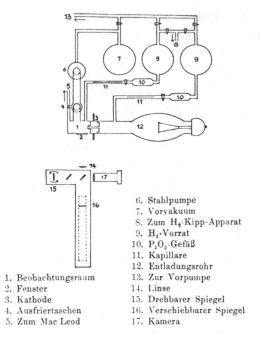

1. Beobachtungsraum
2. Fenster
3. Kathode
4. Ausfriertaschen
5. Zum Mac Leod

6. Stahlpumpe
7. Vorvakuum
8. Zum H₂-Kipp-Apparat
9. H₂-Vorrat
10. P₂O₅-Gefäß
11. Kapillare
12. Entladungsrohr
13. Zur Vorpumpe
14. Linse
15. Drehbarer Spiegel
16. Verschiebbarer Spiegel
17. Kamera

Figure 4.8. One of the world's leading groups for canal ray research was at the University of Munich, headed by Wilhelm Wien. His graduate student Harald Straub was given the task of testing Rupp's claims; after Wien's death in 1928, Straub's work was supervised by Walther Gerlach and Eduard Rüchardt. The above diagram depicts Straub's setup for the rotated mirror experiment. His experiments clearly contradicted Rupp's claims: the latter had been able to measure up a precise angle that exactly confirmed Einstein's value for the mirror rotation, but Straub could not get any interference for any position of the rotating mirror. Furthermore, if he removed lens 14, which took away the need to rotate the mirror, Straub could not find any interference at all for hydrogen canal ray sources. However, in that case Rupp had found interference up to a path difference of 15.2 cm. With mercury canal ray sources, and without lens 14, Straub had found interference at a path difference of at most 0.7 mm, whereas Rupp had claimed to have seen interference at as much as 62 cm. This value was just under the maximally attained value with stationary mercury sources – but experts, such as the British spectroscopist Robert Atkinson, had expected that the formation of a canal ray beam would greatly reduce the maximum coherence length of the emitted light. From Straub (1930), p. 648.

Laue that the canal ray experiments had been formulated as "cases in which our [theoretical] knowledge would make a decision possible, even without carrying out an experiment."[31] Apparently, experimentation had devalued yet more in Einstein's system.

---

[31] "[...] Falle, in welchem unser Wissen eine Entscheidung ermöglichte, sogar ohne Ausführung eines Experimentes." Einstein to Max von Laue, 29 August 1936, EA 16 113.

4.2 Experience in Einstein's philosophy

Historian of science Klaus Hentschel has looked at Einstein's relation to experiment with a particular focus on tests of relativity. Hentschel too has observed a shift:[32] Einstein initially showed great interest in the promotion and outcome of experimental tests of, in particular, general relativity. But after the theory had found its most pronounced confirmation through the 1919 English eclipse expeditions, his interest gradually waned. The consistency in the network of arguments and experiments that supported the theory became so convincing that Einstein began to dismiss the possibility of contrary results, even if for a long time he was willing to acknowledge that they would, should they hold up when subjected to closer scrutiny, undermine his theory (he would not consider accommodating such results by introducing ad hoc hypotheses).

Near the end of his life Einstein expressed the extreme position that "[i]f there were absolutely no light deflection, no perihelion motion and no redshift, the gravitational equations would still be convincing because they avoid the inertial system. [...] It is really quite strange that humans are usually deaf towards the strongest arguments, while they are constantly inclined to overestimate the accuracy of measurement."[33] Einstein's disregard for experiment eventually thus took on an almost programmatic tinge: he wanted to impress upon physicists and philosophers that the role of experiment should not be overrated, as he apparently by now had come to feel it often was.

This more ideological aspect of the older Einstein's position on experiment is also reflected in his later statements on the role of the Michelson–Morley experiment in the formulation of the special theory.[34] To clarify: it has long been argued that the role of this experiment in Einstein's creation of special relativity has been rather small and indirect, in contrast to what the established lore initially held. That lore, which in its stronger versions would assert that the special theory of relativity had been found through "a generalization from Michelson's experiment," had been promulgated by experimentalists such as Robert A. Millikan and empiricist philosophers such as Hans Reichenbach.[35] This viewpoint has been largely dispelled, in particular by reference to pronouncements of Einstein from the 1940s and 1950s that threw doubt on his knowledge of the Michelson–Morley experiment prior to 1905; generally, Einstein would argue that if he had known about the experiment at all, it still would not have exerted a great influence on him since he had already become convinced of the principle of relativity before he might have

---

[32] See Hentschel (1992), for example pp. 615–616.
[33] Einstein to Max Born, 12 May 1952, as quoted in Hentschel (1992), p. 615.
[34] On Einstein and the Michelson–Morley experiment, see van Dongen (2009), and references therein.
[35] Quoted is Robert Millikan (1949), p. 343; for discussion, see Holton (1969), on pp. 293–298 in Holton (1988).

learnt of Michelson's experiment.[36] Yet, surprisingly, a recently uncovered lecture that Einstein gave in Chicago in 1921 suggests otherwise: after first stating that at a young age he had tried to think of experiments that would exhibit an ether drift, Einstein told his audience that "when I was a student, I saw that experiments of this kind had already been made, in particular by your compatriot, Michelson."[37] His 1921 lecture thus suggests that Einstein learnt of the experiment before he had become convinced of the principle of relativity.

How are we to understand the shift in emphasis awarded the Michelson–Morley experiment? Firstly, it is quite possible that in 1921 Einstein wished to please Michelson's home crowd – Michelson was not present during Einstein's visit, but he was a prominent faculty member at the University of Chicago. Yet, later in life, even on occasions that honored Michelson's career in physics, Einstein would not similarly single out the Michelson–Morley experiment. At an event in 1952 that celebrated the centenary of Michelson's birth, for example, he stated that the experiment had only had a "rather indirect" influence on his work.[38]

The intended audience for such statements, we believe, were overambitious empiricists: immediately after downplaying the role of the Michelson–Morley experiment at the 1952 event, Einstein pointedly continued with his familiar observation that "[t]here is, of course, no logical way leading to the establishment of a theory but only groping constructive attempts."[39] This evidently conflicted with the "prejudice" that he attributed to scholars with a "positivistic philosophical attitude." These held that "facts by themselves can and should yield scientific knowledge without free conceptual construction."[40] His recollections of the role of the Michelson–Morley experiment were intended to prove these positivists wrong.

The last quotation above is taken from Einstein's 1949 "Autobiographical notes," which we have already often consulted. That article's intended audience

---

[36] See for example Einstein's comments to Michelson's biographer, Bernard Jaffe, as quoted on p. 340 in Holton (1988).

[37] "Als junger Mensch hatte mich als Physiker die Frage interessiert, welches das Wesen des Lichtes ist, und im Speziellen, was die Beschaffung des Lichtes zum Koerper ist [sic], naemlich es war so, als Kind schon lehrte man mich, dass das Licht den Schwingungen des Lichtaethers untersteht. Wenn das der Fall ist, so muss man das merken koennen, und so dachte ich darueber nach, ob es moeglich sei, durch irgendwelche Experimente zu merken, dass die Erde im Lichtaether bewegt ist. Aber als ich im Studium war, da sah ich, dass Experimente dieser Art schon gemacht worden waren, insbesondere von Ihrem Landsmann, Michelson." Einstein in "Vortrag des Herrn Prof. Dr. Albert Einstein, Chicago, 5. Mai. 1921, Francis Parker School"; see Appendix D in Buchwald *et al.* (2009). For analysis, including that of similar statements by Einstein possibly made in Kyoto in 1922, see van Dongen (2009).

[38] Einstein as quoted in Holton (1969), on p. 303 in Holton (1988); see also Shankland (1964). Einstein contended that the Michelson–Morley experiment had only made clear to him the unnatural character of Lorentz's theory; on this occasion he pointed to his conviction that the electromotive force induced in bodies moving in a magnetic field was nothing but an electric field as a more direct influence.

[39] Einstein as quoted in Holton (1969), on p. 303 in Holton (1988); see also Shankland (1964).

[40] As in Einstein (1949a), p. 49.

also consisted of empiricist physicists and philosophers (the volume in which it appeared was thus appropriately called *Albert Einstein: Philosopher-Scientist*), and it similarly de-emphasized the role of experiment. For example, Einstein claimed there that the "fundamental crisis" that had been laid bare by Max Planck's quantum had been "remarkable," for "at least in its first phase, it was not in any way influenced by any surprising discoveries of an experimental nature."[41] Einstein did not specify exactly what he understood to be the quantum's "first phase," but anyone familiar with the history of the black body problem will be surprised by this representation of events: one need only think of the experiments conducted at the *Physikalisch-Technische Reichsanstalt* by men such as Wilhelm Wien and Heinrich Rubens to realize that Einstein was simply wrong.[42] Einstein himself, too, knew better, for when he actually described the developments leading up to Planck's black body result, he did in fact include experimental contributions in his account.[43] Thus, the above remark should really be seen as a programmatically motivated statement, included to reinforce the autobiography's larger philosophical claim. That larger claim, as we have seen earlier, was that true advancement and understanding in physics can and will be achieved by sticking to the unified field theory methodology. Downplaying the role of experiment and taking on empiricist philosophers served to promote that methodology and, thereby, also the practice of classical field theory.

Einstein's autobiography thus brings us back to his philosophical writings and the place awarded experience there – yet another angle from which to describe his epistemological reorientation: the role of experience did not just change in Einstein's practice in physics, it also changed in his philosophy. An early indication of this change was of course the replacement of induction with intuition in his analysis of scientific creativity, as we saw him do in 1919 in his *Berliner Tageblatt* article.[44] The dividing line between induction and intuition is obviously difficult to draw in a general and exact way, yet Einstein's critique of induction expressed that reasoning from the phenomena had begun to take a less prominent and more insecure role in his epistemology.

Nevertheless, in 1918 Einstein was still "angered" by a suggestion that the general theory of relativity owed its formulation to speculation rather than experience: "I believe that this development teaches [...] the opposite, namely that for a theory to deserve trust, it has to be built on generalizable *facts*." To drive his point home he concluded that "a truly useful and deep theory has never been found in a purely

---

[41] As in Einstein (1949a), p. 37.
[42] For the history of the black body law, see for example the recent review article by Gearheart (2002); for particular references to early experiments, see Kuhn (1978), pp. 3–11, Jammer (1989), chapter 1.
[43] See for example p. 39 in Einstein (1949a).
[44] For another example of the same argument, see Einstein (2008), p. 457.

speculative fashion."[45] Similarly, as we saw, in 1919 Einstein found relativity theory to be a "principle" theory: deductive in structure, but the central principle of equivalence was still grounded in a "fact of experience," namely the equivalence of inertial and gravitational mass.[46] The question remains, thus, how the change in Einstein's philosophical assessment of experience interacted with his recollections of 1915. And what can we conclude about the role of both in his involvement in the unified field theory program?

In a 1921 letter to Hendrik A. Lorentz, Einstein lamented a lack of *physical clues* for attempts at unification – these he still believed to be the only possible starting point for "real progress."[47] However, a few years later, in his Gothenburg Nobel lecture delivered in 1923, Einstein reluctantly admitted that mathematics was guiding his current work in unified field theory: "Unfortunately we cannot in these efforts base ourselves on empirical facts as in the derivation of the theory of gravity (equality of gravitational and inert mass), but we are limited to the criterion of mathematical simplicity, which is not free from arbitrary aspects."[48] Yet later, in Argentina in 1925, he would on the basis of his familiar critique of induction (to be replaced by intuition) deny a substantial role for experience in the production of theory. He now made the bold statement that "experience is [...] the judge, but not the generator of fundamental laws" – although he still regarded "empirical understanding of the field in question" to be a *sine qua non* for the creative scientist.[49]

Little by little Einstein nudged his field theory epistemology away from experience and experiment. In 1929, when promoting his unified teleparallel geometry, the field equations of general relativity were presented as the successful result of a "purely mental process," grounded in the "conviction of the formal simplicity of the structure of reality." That success now emboldened Einstein to "proceed along this speculative road, the dangers of which everyone who dares to follow it

---

[45] Einstein to Michele Besso, 28 August 1918: "In Deinem letzten Brief finde ich beim nochmaligen Lesen etwas, das mich geradezu erbost: die Spekulation habe sich als der Empirie überlegen gezeigt. Du denkst dabei an die Entwicklung der Relativitätstheorie. Aber ich finde, dass diese Entwicklung etwas anderes lehrt, das fast das Gegenteil davon ist, nämlich, dass eine Theorie, um Vertrauen zu verdienen, auf verallgemeinerungsfähige *Thatsachen* aufgebaut sein muss. [...] Niemals ist eine wirklich brauchbare und tiefgehende Theorie wirklich rein spekulativ gefunden worden." Doc. 607, pp. 864–865 in Schulmann *et al.* (1998), emphasis as in original; for a related discussion, see also pp. 559–560 in Fölsing (1998).

[46] See Einstein (1919b), on p. 251 in Einstein (1994).

[47] "Wirkliche Fortschritte kann man eben doch wohl nur auf Grund hinreichender *physikalischer* Anhaltspunkte machen; an solchen fehlt es aber leider." Einstein to Hendrik A. Lorentz, 30 June 1921, EA 16 541; emphasis in original.

[48] "Leider können wir uns bei dieser Bemühung nicht auf empirische Tatsachen stützen wie bei der Ableitung der Gravitationstheorie (Gleichheit der trägen und schweren Masse), sondern wir sind auf das Kriterium der mathematischen Einfachheit beschränkt, das von Willkür nicht frei ist." As in Einstein (1923a), p. 9.

[49] See Einstein (2008), pp. 451–452.

should permanently keep before his eyes."[50] Another four years later, Einstein no longer spoke of dangers and instead confidently stated in his Spencer lecture that "the creative principle resides in mathematics." This of course combined well with the notion that experience is not the generator of fundamental laws – in Oxford, he lectured that experience may "suggest concepts" but was otherwise limited to testing theory.[51]

The distance to experience would remain in Einstein's philosophy in the years that followed: in 1950, he called the motivation to formulate new theories because of a conflict with observation in the old ones "trivial, imposed from without." He believed that unification and simplification of the premises constituted a "more subtle motive."[52] Einstein felt that one needed "to apply free speculation to a much greater extent than is presently assumed by most physicists."[53] The downside would nonetheless be that as the creation of theory moved further away from experience, so did the ability to produce testable predictions: "In the generalized theory [of the 'total' field] the procedure of deriving from the premises of the theory conclusions that can be confronted with empirical data is so difficult that so far no such result has been obtained."

As we saw at the end of Chapter 1, in the autobiography the increased distance from experience had acquired a motivation through Einstein's recollections of 1915 in the most explicit fashion. We have just seen that those recollections had clearly undergone a change: in 1918 he had held that general relativity had been built on "facts," but in the Autobiographical notes we find that the same theory of general relativity had taught him that there was "no way" to construct a theory starting from a "collection of empirical facts," no matter how exhaustive that collection may be.[54]

It seems fair to conclude that Einstein's recollection of 1915 was not just directing his work to the study of unified fields – the above has shown that, in turn, his involvement with unified field theory also re-shaped his recollections of the route to general relativity. This suggests further that his methodological positions were likewise just as much influenced by his practice of field theory, as that they gave direction to the latter. Of course, promoting an epistemology in which maxims of

---

[50] "Das Gelingen dieses Versuches aus der Überzeugung der formalen Einfachheit der Struktur der Wirklichkeit heraus auf rein gedanklichem Wege subtile Naturgesetze abzuleiten, ermutigt zu einem Fortschreiten auf diesem spekulativen Wege, dessen Gefahren sich jeder lebhaft vor Augen halten muss, der ihn zu beschreiten wagt." On p. 127 in (Einstein, 1929); see also pp. 559–560 in (Fölsing, 1998).

[51] As in Einstein (1933a), p. 300 in Einstein (1994).

[52] As in Einstein (1950), p. 13.

[53] As in Einstein (1950), p. 16.

[54] See Einstein (1949a), p. 89. Unification was similarly motivated in for instance Einstein (1950), at p. 15: "It has to be admitted that general relativity has gone further than previous physical theories in relinquishing 'closeness to experience' of fundamental concepts in order to attain logical simplicity."

mathematical naturalness and logical simplicity play an essential role also helped to promote his unification attempts. Similarly, a recollection of 1915 in which mathematics had been the quintessential creative element would further strongly serve to promote his field theory program.

Various commentators have identified a more or less exact point in time at which Einstein turned around his epistemology. They have however pointed to different moments at which this change took effect. Pais and Hentschel believe that Einstein's attitude changed around 1927; another biographer however points to 1923 and yet another historian to 1921 as the turn-around years. Each present their own reasons for calling attention to exactly these years – the dates have been motivated by references to, respectively, Einstein's stepped up production of unification attempts, the influence of Émile Meyerson's philosophy, Einstein's 1923 Gothenburg lecture and finally Einstein's involvement with Kaluza's unified field theory in 1921.[55] We agree that the most pronounced changes took place during, roughly, the 1920s. Yet, the overall picture that emerges when surveying the larger period is not that there was a single moment that produced a sudden about-face; rather, we observe a gradual process in which Einstein's practice in physics grew ever more engrained with unified field theory attempts while it engaged less with actual experimentation, just as experience and experiment became less and less central in his philosophy of physics. Einstein's recollections of his route to the general relativistic field equations altered in tandem with this development, while they became the most important source that he pointed to to motivate his choices.

Asking which influenced which – the experience of 1915, the unified field theory attempts and the withdrawal from experiment   is like asking a chicken and egg question. Einstein's movement towards mathematical unification does seem to have re-shaped his recollections of his route to general relativity, while it led him away from experiment. Yet, conversely, we have seen that the experience and recollection of 1915 would also motivate Einstein's increased involvement with unified field theory. Rather than a sudden reversal, we should think of Einstein's development as due to a continuously ongoing process in which these various factors and choices were interacting, steadily shifting the balance between experience and theory, and between the roles awarded physical argument and mathematics.

The positions that were the outcome of this process could sometimes manifest themselves quite explicitly, as in his decision to criticize the quantum program. Changes could also remain implicit, perhaps even subconscious – the example of the changing recollection of 1915 comes to mind. Overall, Einstein's choices led his practice in physics further away from experiment and experience, and his epistemology and ideas about method more and more towards the notion that mathematics

[55] See Pais (1982), p. 329, Hentschel (1992), p. 615, Fölsing (1998), p. 560, Wünsch (2005), p. 296.

is the essential creative element in theory construction – just as his recollections of 1915 increasingly resonated with this development, while they grew in their constitutive and justifying role. By 1933 Einstein had firmly arrived at the position that mathematics was, and in 1915 had been, his true guide; in the next chapter we will present a case study that addresses how that notion interacted with his contemporary field theory research.

5

The method as directive: semivectors

In his 1933 Herbert Spencer lecture on the method of theoretical physics Einstein told his audience that the gravitational field equations can be found by looking for the mathematically simplest equations that a Riemannian metric can satisfy. However, the gravitational field equations were not the only example that Einstein gave in his lecture. His second example was taken from his recent work.

At this point we still lack a theory for those parts of space in which the electrical charge density does not disappear. De Broglie conjectured the existence of a wave field, which served to explain certain quantum properties of matter. Dirac found in the spinors field-magnitudes of a new sort, whose simplest equations enable one to a large extent to deduce the properties of the electron. Subsequently I discovered, in conjunction with my colleague, Dr. Walther Mayer, that these spinors form a special case of a new sort of field, mathematically connected with the four dimensional system, which we called "semivectors."[1]

Einstein, together with Mayer, had in the course of 1932 turned to the Dirac theory of the electron. They had introduced the "semivector": a generalization of the spinor, as the above suggests. Yet, semivectors were not only more general than spinors, they were also simpler:

These semivectors are, after ordinary vectors, the simplest mathematical fields that are possible in a metrical continuum of four dimensions, and it looks as if they describe, in a natural way, certain essential properties of electrical particles.

The diligent mathematical study of these new semivectors had given Einstein and Mayer a most appealing result:

The simplest equations which such semivectors can satisfy furnish a key to the understanding of the existence of two sorts of elementary particles, of different ponderable mass and equal but opposite electrical charge.

---

[1] As in Einstein (1933a), on p. 301 in Einstein (1994).

It is thus not surprising that Einstein would mention the semivector in his Herbert Spencer lecture: his new understanding of the existence of the electron and proton, the "two sorts of elementary particles" he was referring to, could serve as an example par excellence of a fruitful application of his unified field theory methodology.

In this chapter we intend to study Einstein and Mayer's semivectors in more detail to see how Einstein's epistemological thought interacted with his unification program in physics. The question that we will hope to address in particular is whether Einstein's re-shaped methodological convictions were involved in the semivector research. On the basis of the Herbert Spencer lecture one would expect this to have been the case; his conclusion on that occasion was:

The important point for us to observe is that all these constructions [i.e. the semivector, Maxwell and gravitational field] and the laws connecting them can be arrived at by the principle of looking for the mathematically simplest concepts and the link between them. In the limited number of the mathematically existent simple field types, and the simple equations possible between them, lies the theorist's hope of grasping the real in all its depth.

5.1 The unnaturalness of the spinor

Einstein published four papers on semivectors, and they were all co-authored by his assistant Walther Mayer. The first article appeared in the proceedings of the Berlin Academy in November 1932.[2] The next two papers came out in the proceedings of the Royal Academy of Sciences in Amsterdam, and the last short note was published in the Princeton journal *Annals of Mathematics*, early in 1934.[3] The sequence of papers reflects that Einstein was uprooted by events in Germany in 1933 and spent most of that year in transit. The grave public and personal circumstances are also exhibited in the correspondence on spinors and semivectors that Einstein entertained with his close friend Paul Ehrenfest. We will briefly discuss this correspondence on the next pages; a sketch of Walther Mayer and his relation with Einstein is given in the boxed text. We end this section by outlining the goals that Einstein and Mayer had set for the semivector theory.

5.1.1 Ehrenfest and the spinor

"Despite the great importance that the spinor concept, introduced by Pauli and Dirac, has taken on in molecular physics, it cannot be contended that the present mathematical analysis of this concept complies with all justified demands. This is

---

[2] See Einstein and Mayer (1932b). The starting point of the work on semivectors is documented in a letter by Mayer, dated 27 August 1932, when he writes to Einstein that he is most eager to learn the latter's new ideas regarding the Dirac equation; W. Mayer to Einstein, 27 August 1932, EA 18-137.
[3] See Einstein and Mayer (1933a,b, 1934).

the reason why P. Ehrenfest has impressed upon one of us with great vigor that we should make an effort to fill this gap."[4] Thus begins Einstein and Mayer's first article on semivectors. They were, however, not the only ones to have been prodded by Paul Ehrenfest to study spinor theory. Ehrenfest had introduced the term "spinor," and earlier, in 1929, he had asked Dutch mathematician Bartel van der Waerden if it would be possible to formulate a spinor analysis that was just as straightforward to learn and apply as tensor analysis. Van der Waerden soon published a paper entitled "Spinoranalyse" that attempted to do just that. His approach is still appreciated as an insightful treatment of spinors and we will use the representation given in his work in our next section.[5] Ehrenfest, however, was still not satisfied after van der Waerden's article appeared. He felt that the spinor lacked a simple, intuitive interpretation and in the summer of 1932, he convinced Einstein of the same.[6]

In that same year Ehrenfest wrote a short article in the *Zeitschrift für Physik*, the leading publication for researchers in quantum theory, modestly entitled "Einige die Quantenmechanik betreffende Erkundigungsfragen" – "Some inquisitive questions regarding quantum mechanics."[7] He animatedly repeated the question that he had put to van der Waerden before, along with a number of other questions and appeals – quite in the spirit of what his biographer, Martin J. Klein, has identified as a commitment to "clarity and intelligibility in the flood of new developments" characteristic of Ehrenfest.[8] Ehrenfest did acknowledge the advances made by van der Waerden, but he still felt that one lacked "a *slim booklet* from which one can *pleasantly* acquire spinor calculus and its relation to tensor calculus."[9] Ehrenfest hoped for a publication that would elucidate the representations of the Lorentz group for arbitrary dimensions and he expressed his discomfort with the spinor's lack of *Anschaulichkeit*, or, that is, visualizability. A Weyl spinor has of course two internal components instead of four spacetime components, and, intuitively speaking, somewhat peculiar transformation properties since under a spatial rotation over 2π radians, it does not go over into itself but rather ends up pointing in the opposite direction.

All was not well with Ehrenfest. His marriage to Tatiana Afanassjewa, a strong intellectual in her own right, had become increasingly troubled.[10] He further felt

---

[4] "Bei der grossen Bedeutung, welche der von Pauli und Dirac eingeführte Spinor-Begriff in der Molekularphysik erlangt hat, kann doch nicht behauptet werden, dass die bisherige mathematische Analyse dieses Begriffes allen berechtigten Ansprüchen genüge. Dem ist es zuzuschreiben, dass P. Ehrenfest bei dem einen von uns mit grosser Energie darauf gedrungen hat, wir sollten uns bemühen, diese Lücke auszufüllen." As in Einstein and Mayer (1932b), p. 522.

[5] See van der Waerden (1929); see also van der Waerden (1960), p. 236, for mention of Ehrenfest's role; for positive commentary, see for example Penrose and Rindler (1984), p. vii.

[6] According to a later letter of Einstein to Ehrenfest, 19 May 1933, EA 10-253.

[7] See Ehrenfest (1932). Surprisingly, this article is not cited in Einstein's papers on semivectors.

[8] See Klein (1970), p. xv.

[9] "[Es] fehlt ein *dünnes Büchlein*, aus dem man *gemütlich* die Spinorrechnung mit der Tensorrechnung vereinigt lernen könnte." See Ehrenfest (1932), p. 558, emphasis as in original.

[10] On this see for example Casimir (1983), pp. 174–177.

that he could no longer keep up with the new direction physics had taken and this deeply grieved him. Ehrenfest had made important contributions to the development of quantum theory (for instance by introducing the "adiabatic principle") but now he believed that the theory had become too mathematical, and to him its physical interpretation had become obscured. The tone of his paper was at the same time apologetic, upset, inquisitive and polemical; he feared that his questions might be pointless, yet he hoped for a poignant reply that even "provincial schoolmasters"[11] such as himself could understand. Ehrenfest had lingered with the publication of his short note for about a year, fearing the scorn or – perhaps even more painful – an all too courteous response from Pauli or Bohr. Wolfgang Pauli indeed replied in a private letter to Ehrenfest, expressing however great delight in the latter's piece. He addressed Ehrenfest's points with great care, but left the call for an insightful treatment of the spacetime transformation properties of spinors to someone else: he would rather read that booklet than write it. On Ehrenfest's instigation, the contents of Pauli's letter were soon published.[12]

Ehrenfest found such expository publications badly needed to facilitate the instruction of the new theory to students; he felt it increasingly difficult to teach the quantum theory. Ehrenfest had succeeded Lorentz in Leiden, yet he believed that he could no longer bear the responsibility of this chair. Earlier in 1932, he had asked Einstein to seek a less demanding position for him in the United States, but to no avail.[13] Ehrenfest had now – bereft of the joy that physics once gave him – truly fallen into despair and on 14 August 1932 he wrote an alarming letter to his friends and colleagues, among them Einstein.[14] In the letter, he expressed his deep desperation and considered the possibility of suicide. Ehrenfest's next letter to Einstein, sent five days later and three days after submitting his "Erkundigungsfragen," contained extensive notes on the spinor representations of the Lorentz group, copied from a book by Hermann Weyl.[15] Ehrenfest's deep distress again surfaced as he exclaimed over and over his failure to understand what he was copying. In another letter to Einstein, sent just two days later, Ehrenfest conveyed his hope that "I will finally understand through *your* exposition how everything works out."[16]

Their correspondence picked up again well after Einstein's first paper on semivectors had appeared.[17] In the article, after acknowledging Ehrenfest's role (perhaps Einstein had been more inclined to take up the subject partly because of a hope

---

[11] "[P]rovinziale Schulmeister," Ehrenfest to Pauli, 31 October 1932, Doc. 295 in von Meyenn (1985), p. 136.

[12] For the Pauli–Ehrenfest correspondence, see von Meyenn (1985); Pauli's paper is Pauli (1933a).

[13] See Einstein to Ehrenfest, 3 April 1932, EA 10-229.

[14] P. Ehrenfest to Niels Bohr, Einstein, James Franck, Gustav Herglotz, Abram Joffe, Philipp Kohnstamm and Richard Tolman, 14 August 1932, EA 10-236.

[15] P. Ehrenfest to Einstein, 19 August 1932, EA 10-237; Weyl's book is Weyl (1931b).

[16] "[I]ch schliesslich aus *Deiner* Darstellung begreifen werde, wie alles ineinander sitzt." Ehrenfest to Einstein, 21 August 1932, EA 10-242.

[17] That is, as their correspondence is to be found in the Einstein Archive and the Paul Ehrenfest Archive in Leiden (for the latter, see its catalog compiled by Bruce Wheaton (Communication 151 of Museum Boerhaave, Leiden, 1977)).

thereby to diminish his friend's destructive sense of superfluousness), Einstein and Mayer continued their introduction of semivectors by claiming that:

Our efforts have led to a derivation that in our opinion complies with all demands for clarity and naturalness and completely avoids obscure tricks. In doing so – as will be shown in what follows – the introduction of new quantities, the "semivectors," has proven to be necessary. These incorporate spinors, but essentially have a more transparent transformation character than the spinors.[18]

The semivectors were an honest attempt to answer to Ehrenfest's appeal: their foremost advantage was that they were mathematically more natural than spinors. One senses from the above that Einstein agreed with Ehrenfest that in particular its transformation properties made the spinor *unanschaulich*. Nevertheless, Einstein did not make clear exactly why the transformation properties of the semivector were more transparent than the those of the spinor. In any case, for reasons of transparency, and to elucidate the relations between spinors, spacetime and tensors, Einstein forwarded the semivector. Einstein and Mayer's next paper contained a footnote, acknowledging Ehrenfest's insistence on the importance of searching for a "logically simple and transparent analysis of the spinors."[19]

In the end, the biggest success that Einstein believed was achieved through the introduction of semivectors was the explanation that they could give of the occurrence of oppositely charged particles with differing masses: electrons and protons. As he told his Oxford audience, this success was the reward for asking for the simplest possible equations that the mathematically natural semivectors could satisfy. Ehrenfest was entrusted with the publication of this paper:

Enclosed is a paper by myself and Mayer, which can also be seen as a result of the spirited stimulus that you have given me last summer. I believe that it truly contains a way to understand the electron and the proton. The mathematical theory is now of an exemplary simplicity. Please be so kind as to submit the paper to the Dutch Academy for publication in its proceedings. I would like the proofs to be sent to professor Walther Mayer, villa Mon Repos, Le Coq, and if this should be before the 20th of June, also a copy to me in Oxford, Christ Church College.[20]

---

[18] "Unsere Bemühungen haben zu einer Ableitung geführt, welche nach unserer Meinung allen Ansprüchen an Klarheit und Natürlichkeit entspricht und undurchsichtige Kunstgriffe völlig vermeidet. Dabei hat sich – wie im folgenden gezeigt wird – die Einführung neuartiger Grössen, der „Semi-Vektoren", als notwendig erwiesen, welche die Spinoren in sich begreifen, aber einen wesentlich durchsichtigeren Transformationscharakter besitzen als die Spinoren" (Einstein and Mayer, 1932b, p. 522).

[19] "[L]ogische einfache und durchsichtige Analyse der Spinoren," (Einstein and Mayer, 1933a, p. 497).

[20] "Beiliegend erhältst du eine Arbeit von mir und Mayer, die auch noch als eine Folge der temperamentvollen Anregung zu betrachten ist, die Du mir im letzten Sommer gegeben hast. Ich glaube, sie enthält wirklich ein Mittel zum Verständnis des Elektrons und Protons. Die mathematische Theorie ist nun von vorbildlicher Einfachheit. Sei so freundlich und übergib die Arbeit der holländischen Akademie zum Druck in ihren Sitzungsberichten. Die Korrektur bitte ich an Professor Walther Mayer, Villa Mon Repos, Le Coq zu senden und, wenn dies schon vor dem 20. Juni sein sollte, auch ein Exemplar an mich nach Oxford, Christchurch College." Einstein to P. Ehrenfest 19 May 1933, EA 10-253.

Walther Mayer: collaborator, yet assistant

In 1929 Einstein was looking for a new assistant. He consulted the mathematical physicist Richard von Mises in Vienna and von Mises brought Walther Mayer (1887–1948; see Figure 5.1) to Einstein's attention.[21] Mayer was at the time *Privatdozent* in Vienna – he had also obtained his doctorate in Vienna, and had studied at the Zurich ETH, Einstein's old alma mater. Mayer's expertise was in differential geometry and topology.

In his very first letter to Mayer, Einstein indicated that he intended to continue working on the teleparallel unified field theory. He informed Mayer that his colleagues did not think highly of this theory and wanted to know if Mayer nevertheless did believe in it, as any collaboration would otherwise be pointless.[22] Mayer had no such qualms and the collaboration immediately proved to be productive, as is documented by a swiftly published joint paper.[23]

Einstein came to appreciate highly Mayer's contributions and when he set out to obtain funding to continue the latter's position in Berlin, he did not stop short of suggesting that if this posed any problems he might be inclined to look favorably at a recent offer made to him by the California Institute of Technology in Pasadena – "the continued collaboration with this man" had become no less than "a question of life and death"[24] (see Figure 5.2). Eventually funding for Mayer was found, first in Germany and later through the American Macy Foundation.

Einstein's tumultuous existence in 1933 greatly hindered his collaboration with Mayer on the Dirac equation and semivectors. Einstein often complained that he could hardly find the time for much desired diligent work.[25] Early in 1933 Mayer was in Vienna and had intended to go to Berlin, but under the given circumstances he had to change his plans, and in fact he would have to do so time and again; more than once a rendez-vous was hampered.[26]

The second and third paper on semivectors appeared in Amsterdam because, following the National Socialist rise to power, Einstein had resigned from the Berlin Academy.[27] At the time he was visiting the United States and there he had publicly spoken out strongly against the new regime. He had vowed not to go back to Germany, in part because he feared for his own personal safety. When he returned

---

[21] Richard von Mises to Einstein, 17 December 1929, EA 18-225.
[22] Einstein to Mayer, 1 January 1930, EA 18-065.
[23] See Einstein and Mayer (1930).
[24] "[D]ie dauernde Zusammenarbeit mit diesem Manne" [...] "eine Lebensfrage;" Einstein to Ludwig Bieberbach (Prussian Academy of Sciences), 19 June 1930, EA 18-227.
[25] Einstein to Mayer, 23 February 1933, EA 18-163.
[26] Einstein to Mayer, 25 March 1933, EA 18-164.
[27] For more on Eistein's departure from Berlin, see for example Stern (1999), pp. 152–154, Levenson (2003), pp. 417–421, Isaacson (2007), pp. 401–424; Rowe and Schulmann (2007), pp. 266–282.

When Einstein decided to start working in the United States, he also wanted to safeguard his collaboration with Mayer. This put him in a difficult position with both Mayer and the Institute for Advanced Study in Princeton, his new employer. Mayer felt that he would have to swallow a demotion if he were to go to Princeton, and Einstein suggested to the Institute that Mayer might rather go to Spain instead, greatly harming their joint work.[28]

It seems that Mayer was too reserved to further his case himself. He mostly relied on Einstein's movements on the one hand, yet all the same complained that he was being regarded as the latter's "appendix."[29] From this episode Mayer comes across as a bit of a worry-laden character, held back by various inhibitions. He repeatedly needed to be spurred on by Einstein; for example, Einstein suggested that Mayer should give a seminar in the Princeton mathematics department to ease the issue regarding the latter's tenure. Anticipating that Mayer's reply was likely to be a bit reluctant, he added that no one would honor him (Einstein) with such a heavy responsibility.[30]

Mayer was disappointed that his work was not fully appreciated and felt that he was unwelcome at the Institute. In the end he nevertheless did decide to go to Princeton and he and his wife left for the United States in Einstein's entourage. Mayer became an associate professor at the Institute, with tenure. Once there, his collaboration with Einstein ended surprisingly soon.[31] Mayer returned to his old trade, topology. The Einsteins and Mayers nevertheless remained friends after the professional dissociation. In a 1948 letter to Einstein's sister Maja, Mayer expressed that he still enjoyed his work but that he had found it difficult to acclimatize to the United States. He and his wife had trouble with English, yet also believed that their German was waning; they felt lonely.[32] Later that year Walther Mayer suffered an untimely death from cancer.

to Europe he stayed on the Belgian coast at De Haan (or, in French, Le Coq sur Mer), later joined by Walther Mayer and both their families – under the protection of two guards, assigned by the Belgian government.

Einstein made a number of visits on the European continent – not into Germany however – and traveled twice to the United Kingdom. On one of these occasions he gave the Herbert Spencer lecture in Oxford. Three months later, on 9 September 1933, Einstein left Europe again, and this time for good. On 17 October he arrived in New York, and from there he went on to Princeton, to take up his position as professor at the Institute for Advanced Study.

---

[28] See Pais (1982), p. 493.

[29] "Anhängsel," as in Einstein to Mayer, 28 January 1933, EA 18-158.

[30] Einstein to Mayer, 28 January 1933, EA 18-158.

[31] According to the recollections of Abraham Pais (1982), pp. 493–494, this was due to Mayer's wish no longer to work on classical or unified field theories, and the belief that his career would be best furthered if he were to pursue independent work. Unfortunately, we have not found material in the Einstein–Mayer correspondence to support this account independently.

[32] Walther Mayer to Maja Einstein, 23 June 1948, EA 18-237.

As a consequence of the new political situation in Germany, Einstein's correspondence with Ehrenfest took a different turn. The German situation exacerbated Ehrenfest's despair. Both Ehrenfest and Einstein took it upon themselves to find a refuge for many of the intellectuals and students who needed to get out of Germany. The newly founded Hebrew University in Jerusalem could have provided a welcome solution, except that it had scarce resources, and Einstein was highly critical of its management and academic quality.[33] Einstein and Ehrenfest were involved in an attempt to organize a university for refugees, preferably in Britain, but the attempt failed.[34] Soon Ehrenfest could no longer see a way out of his deep depression and, on 25 September 1933, he took his own life.

5.1.2 Semivectors: Einstein's goals

Throughout all the personal and political turmoil, Einstein was working on spinors and the Dirac equation. Before we present his work in a more detailed way, we first give a brief outline of his initial goals, as we found them expressed in the published work or in the correspondence with Walther Mayer.

- Spinor-analysis: by introducing semivectors, Einstein wanted to arrive at a mathematically simpler and more general formulation of spinors.
- Incorporating the Dirac equation (with the electromagnetic gauge field) in the general theory of relativity. On a few occasions, Einstein expressed his hope that this would be achieved in the spirit of unified field theories; this meant that the fields should not be arbitrarily added up, but should emanate from some unified Lagrangian.[35]
- Mass term in the Dirac equation: the Schrödinger equation has no mass term, both the Klein–Gordon equation and Dirac equation in general do have a mass term. Initially Einstein wanted to see whether the rest mass could have an electromagnetic origin (e.g. be given by a particular choice for the electromagnetic potential). This problem partly inspired the search for the unified description of electrical particles.[36]

It is important to observe that Einstein did not yield to quantum mechanics during the semivector interlude. He studied the Dirac theory exclusively as a classical field theory; he usually did not write an $\hbar$ in his equations and the theory was in no way quantized beforehand. This implied that all the observables in his theory would have a continuous spectrum, unless the field equations and boundary conditions

---

[33] On the early years of physics at the Hebrew University, see Unna (2000); see also Einstein to P. Ehrenfest, 19 May 1933, EA 10-255.
[34] See for example Einstein to P. Ehrenfest, 9 July 1933, EA 10-262. Einstein also corresponded with the Oxford physical chemist Frederick A. Lindemann on this issue. Lindemann had been trying to win Einstein for Oxford and the Herbert Spencer lecture can be seen as a consequence of Einstein's contacts with Lindemann. For more on Lindemann's role in the relocation in Britain of German scientists, see Wolff (2000).
[35] See for example Einstein to Mayer, 23 March 1933, EA 18-163.
[36] This is particularly clear in a letter from Einstein to Mayer, dated 21 December 1932, EA 18-149.

Figure 5.1. Walther Mayer in 1940 (by permission of the United States Holocaust Memorial Museum, Washington, DC).

constrained the solutions in such a way that they would give discrete values for the observables. Let us look, for instance, at the definition of the electric charge in classical Dirac theory:

$$Q = e \int d^3x \, \psi^\dagger \cdot \psi. \tag{5.1}$$

The ψ are not operators that act on some state, neither are they interpreted as probability distributions. They are ordinary classical fields, like the fields in general relativity. This means that if ψ is given by a superposition of solutions of the Dirac system, measurement does not reduce ψ to one particular solution. In such a classical field theory, there is no difference between the "expectation value" of an observable and the value found upon observation – the value assigned to observables is at any time the outcome of integrals such as (5.1). Finally, the ψ are not normalized; their scale is not fixed and this generally implies that the spectrum of an observable like Q is continuous, even if the spectrum of eigenfunction solutions of the field equations is discrete (the electron charge e has been added for dimensional reasons). Yet, from the semivector Dirac theory Einstein wanted to *derive*:

- Quantization: Einstein hoped to derive particular (non-singular) solutions to the combination of Dirac and Maxwell field equations that would give a discretely valued charge. Despite the fact that Einstein's published equations did not explicitly contain an $\hbar$, he at

Figure 5.2. Einstein and Mayer at Caltech, early 1930s (courtesy of the Institute Archives, California Institute of Technology).

one point believed scale invariance to be broken in the theory and that it could therefore possibly contain an explanation for the atomicity observed in nature.[37] He also expressed the hope to Mayer that he might be able to derive the value of the fine-structure constant from the theory.[38]

The goal of formulating a more natural and general spinor analysis was Einstein's starting point. Its realization was reported in the first semivector paper, which also produced generally covariant semivector Dirac equations.[39] The paper was presented on 10 November 1932, and opened up the possibility to start addressing the other issues. The most important test for the semivector formalism was eventually the last point, whether it yielded solutions of the field equations that gave quantized charges. However, the biggest success that Einstein believed to have derived from the semivector formalism was:

- Unification of charged particles: the semivector generalization of the Dirac equation was thought to give a unified description of charged elementary particles, in particular of the electron and proton.

[37] See in particular Einstein to Mayer, 4 June 1933, EA 18-176.
[38] Einstein to Mayer, 21 December 1932, EA 18-149.
[39] See Einstein and Mayer (1932b).

This result was reported in the two papers that appeared in Amsterdam on 27 May and 24 June 1933. It is also the result that Einstein was so enthusiastic about in Oxford. Gradually however, resistance to the semivector grew. Einstein and Mayer's last paper on semivectors came out in 1934 (it had been submitted earlier, on 8 November 1933). This article was no longer involved with any of the goals above, except for the incorporation of the semivector in general relativity; one senses however that this was presented foremost as an exercise with exclusively mathematical appeal, and of no immediate relevance to Einstein's larger program in physics.

5.2 Semivectors and the unification of charged particles

We now come to the technical analysis of Einstein's semivector work. We will first make some general remarks about the Dirac equation and the spinor representation of the Lorentz group. Then, we will follow Einstein and Mayer in their introduction and elaboration of the semivector formalism. Finally we will try to reconstruct how Einstein's methodological ideas might have interacted with their work.

5.2.1 The Dirac equation and the Lorentz group

The Dirac equation is a wave equation that describes the interaction of charged matter with the electromagnetic fields. The idea that a wave could be associated with material particles was due to Louis de Broglie. He had forwarded it in 1923, motivated by the analogy with dual descriptions of radiation and considerations of Lorentz invariance. If a wave was associated with matter

$$\psi(t,\vec{x}) = \psi_0 \exp(i2\pi(\vec{k} \cdot \vec{x} - \nu t)), \qquad (5.2)$$

then a particle would have to be ascribed a wavelength and a frequency. These were to be given by:

$$k = p/h \qquad \nu = E/h. \qquad (5.3)$$

As an aside, let us point out that Einstein would repeatedly make, in his words, the "De Broglie Ansatz" for his classical fields, in particular in his correspondence; he would not write equations (5.3) in his publications however.

De Broglie's relations did not include electromagnetic interaction terms. This problem appeared to be solved in 1926 with Erwin Schrödinger's famous wave equation, except that the Schrödinger equation was not Lorentz invariant. Its relativistic pendant, today known as the Klein–Gordon equation, was published a little later. Schrödinger had in fact already considered that equation before publishing the non-relativistic equation named after him. He had dismissed it, however, because it did not give the correct results for the fine structure of the hydrogen spectrum. The reason for the discrepancy was that his relativistic wave equation neglected the spin of the electron.

Spin, only recently suggested in 1925, indeed accounted for the discrepancy through its coupling with the electron's orbital angular momentum. As much was known by 1927, but at that time there still was no consistent relativistic theory that incorporated the electron's spin from the beginning. Such a theory was found by Paul Dirac a year later in 1928. It is this theory that Einstein's study of semivectors addressed.[40]

In its familiar form the Dirac equation is written as (with $\hbar = c = 1$):

$$\gamma^k (i\partial_k + eA_k)\psi = m\psi, \tag{5.4}$$

e and m are the particle charge and mass, A^k is the electromagnetic gauge field (in this chapter, Latin indices will run over the four spacetime components). The γ^k are given in the Weyl representation:

$$\gamma^0 = \begin{pmatrix} 0 & -\sigma_0 \\ -\sigma_0 & 0 \end{pmatrix} \qquad \vec{\gamma} = \begin{pmatrix} 0 & \vec{\sigma} \\ -\vec{\sigma} & 0 \end{pmatrix}. \tag{5.5}$$

The Dirac spinor ψ has four components, and it can be given as the sum of two two-component Weyl spinors; σ_k are the Pauli-matrices, with σ_0 the unit-matrix.

The Dirac equation can also be written as a system of two coupled two-dimensional linear differential equations. This gives the *van der Waerden* spinor representation:[41]

$$\begin{cases} \nabla^k \sigma_k'^{\dot{\nu}\lambda} \psi_\lambda = m\psi^{\dot{\nu}}, \\ \nabla^k \sigma_{k\,\lambda\dot{\nu}} \psi^{\dot{\nu}} = m\psi_\lambda \end{cases} \tag{5.6}$$

with $\nabla_k = -i\partial_k + eA_k$, $k = 0, \ldots, 3$; Greek indices, dotted and undotted, run over the two internal spinor components. This representation will prove to be convenient for comparing the spinor and the semivector. The ψ^λ and $\psi^{\dot{\nu}}$ are two two-component Weyl spinors that together make up the Dirac spinor. The σ_k' are defined by $\sigma_0' = \sigma_0$, $\vec{\sigma}' = -\vec{\sigma}$. Spinor indices are raised and lowered by antisymmetric two-by-two matrices with off-diagonal entries $+1$ and -1.

The system has four degrees of freedom; two describe the spin orientation, the remaining two the sign of the mass of the solution – in the quantized theory

---

[40] The original references to Dirac's work are Dirac (1928a,b). For historical introductions, see for example Jammer (1989) and Chapter 1 in Weinberg (1995).
[41] See van der Waerden (1932).

these two distinguish the electron and positron states. The Dirac equation, (5.4) or (5.6), is Lorentz invariant, with the Lorentz transformations acting in the spinor representation. This is its mathematical aspect that Einstein zoomed in on.

The spinor-representation of the Lorentz group is given by the fundamental representation of the group $SL(2,\mathbb{C})$, that is, the group of linear maps of unit determinant that act on a two-dimensional complex vector space.[42] A spinor $\psi^\mu = (\psi^1, \psi^2)$ transforms under an $SL(2,\mathbb{C})$ transformation β^ν_μ, and the conjugate spinor $\psi^{\dot\mu} = (\psi^{\dot 1}, \psi^{\dot 2}) = ((\psi^1)^*, (\psi^2)^*)$ transforms with the matrix $(\beta^*)^{\dot\nu}_{\dot\mu}$.

The following illustrative example shows how the Lorentz group is represented by $SL(2,\mathbb{C})$. An ordinary four-vector $x^k = (t,x,y,z)$ can be written as a two-by-two complex matrix $c_{\mu\dot\lambda}$, by – as above in the Dirac equation – using the Pauli-matrices;

$$c_{\mu\dot\lambda} = x^k \sigma_{k\,\mu\dot\lambda} = \begin{pmatrix} z+t & x-iy \\ x+iy & -z+t \end{pmatrix}. \tag{5.7}$$

Consider now a transformation

$$\begin{aligned} x' &= x & z' &= \tfrac{1}{2}(\eta^2 + \eta^{-2})z + \tfrac{1}{2}(\eta^2 - \eta^{-2})t, \\ y' &= y & t' &= \tfrac{1}{2}(\eta^2 - \eta^{-2})z + \tfrac{1}{2}(\eta^2 + \eta^{-2})t. \end{aligned} \tag{5.8}$$

This corresponds to a Lorentz boost in the z-direction with speed $v = (\eta^4 - 1)/(\eta^4 + 1)$. It is equivalent to transforming the matrix elements $c_{\mu\dot\lambda}$ with

$$\begin{aligned} c'_{1\dot 1} &= \eta^2 c_{1\dot 1} & c'_{1\dot 2} &= c_{1\dot 2} \\ c'_{2\dot 1} &= c_{2\dot 1} & c'_{2\dot 2} &= \eta^{-2} c_{2\dot 2}. \end{aligned} \tag{5.9}$$

This transformation is reproduced by:

$$c'_{\mu\dot\lambda} = (\beta)^\nu_\mu \, c_{\nu\dot\beta} \, (\beta^*)^{\dot\beta}_{\dot\lambda} \tag{5.10}$$

where $\beta = \mathrm{diag}(\eta, \eta^{-1}) \in SL(2,\mathbb{C})$. Boosts in other dimensions can be dealt with in the same way. By using unitary two-by-two matrices with determinant one ($SU(2)$ is a subgroup of $SL(2,\mathbb{C})$), one can similarly construct rotations in the three spatial dimensions. Boosts and rotations, together with space and time inversions, make up all the transformations in the Lorentz group.

To illustrate further the structure of the Lorentz group and its representations, we take up some modern notation. Let the J_i ($i = 1,2,3$) represent the generators of rotations, and the K_i the generators of boosts. These form the Lie algebra:

$$[J_i, J_j] = i\epsilon_{ijk}J_k, \qquad [K_i, K_j] = -i\epsilon_{ijk}J_k, \qquad [J_i, K_j] = i\epsilon_{ijk}K_k. \tag{5.11}$$

---

[42] One should actually take $SL(2,\mathbb{C})/\mathbb{Z}_2$, see Weinberg (1995), pp. 86–91.

By constructing the operators

$$N_i = \frac{1}{2}(J_i + iK_i), \qquad N_i^\dagger = \frac{1}{2}(J_i - iK_i), \tag{5.12}$$

the above algebra can be rewritten as

$$[N_i, N_j] = i\epsilon_{ijk}N_k, \quad [N_i^\dagger, N_j^\dagger] = i\epsilon_{ijk}N_k^\dagger, \quad [N_i^\dagger, N_j] = 0. \tag{5.13}$$

This shows that the Lorentz algebra can be split up into two pieces, each piece generating an $SU(2)$. The two $SU(2)$ in the Lorentz group, however, are not independent. Namely, under space inversion (or "parity") transformations the sign of the generators K_i changes; $K_i \rightarrow -K_i$, so $N_i \rightarrow N_i^\dagger$. This implies that the two $SU(2)$ are interchanged.

The representation of $SU(2)$ is usually labeled by a number $n \in 0, 1/2, 1, 3/2$, etc. For the Lorentz group we thus need two numbers n and m, with obviously also $m \in 0, 1/2, 1, 3/2$, etc. The pair (n,m) labels the representation of the full Lorentz group. Since $J_i = N_i + N_i^\dagger$, we can identify the spin of the representation of the Lorentz group with $n + m$. The Lorentz group representation for the *left-handed* Weyl spinor ψ^α is labeled by $(\frac{1}{2}, 0)$. The representation of the *right-handed* Weyl spinor $\psi^{\dot\alpha}$ is labeled by $(0, \frac{1}{2})$. As follows from the above, these are interchanged under parity. The Dirac spinor $(\frac{1}{2}, 0) \oplus (0, \frac{1}{2})$ is invariant under parity. Finally, from the transformation rule (5.10), we see that a vector transforms as a tensor product of two Weyl spinors, namely as a tensor product of one spinor with an un-dotted index and one spinor with a dotted index. One writes this as $(\frac{1}{2}, \frac{1}{2}) = (\frac{1}{2}, 0) \otimes (0, \frac{1}{2})$.

Einstein and Mayer essentially constructed two alternative "semivector" representations of the Lorentz group. To render their treatment more accessible we will present these here in our modern notation first. In their construction of the semivector representation "of the first kind" they basically used the following set[43] of six generators N_i and iN_i:

$$[N_i, N_j] = i\epsilon_{ijk}N_k, \quad [iN_i, iN_j] = -i\epsilon_{ijk}N_k, \quad [N_i, iN_j] = -\epsilon_{ijk}N_k. \tag{5.14}$$

If we now define $A_i = iN_i$, we retrieve the familiar Lie algebra of the Lorentz group (5.11):

$$[N_i, N_j] = i\epsilon_{ijk}N_k, \quad [A_i, A_j] = -i\epsilon_{ijk}N_k, \quad [N_i, A_j] = i\epsilon_{ijk}A_k. \tag{5.15}$$

Semivectors "of the second kind" transform under the complex conjugate representation; its six generators would then be N_i^* and $-iN_i^*$. Also with these six generators

---

[43] These generators coincide with the fundamental matrices given in equation (14c) of Einstein and Mayer (1932b), p. 526.

the original formulation of the Lie algebra of the Lorentz group can be reconstructed, in a similar way as above. The J_i and K_i have been chosen as Hermitian four-by-four matrices with only purely imaginary elements. This implies

$$N_i^* = \frac{1}{2}(-J_i + iK_i) = -N_i^\dagger. \tag{5.16}$$

It follows from $[N_i, N_j^\dagger] = 0$ that the elements of the Lorentz group in one semivector representation commute with elements in the other.

5.2.2 Introducing the semivector

In their first paper on semivectors Einstein and Mayer started by studying the vector representation of the Lorentz group. An infinitesimal Lorentz transformation on a vector, a_k^i, can be written as

$$a_k^i = \delta_k^i + \epsilon_k^i \tag{5.17}$$

or

$$a_{ik} = g_{ik} + \epsilon_{ik} \tag{5.18}$$

where ϵ_{ik} is infinitesimally small, antisymmetric and real. Einstein and Mayer showed that ϵ_{ik} can be decomposed into two matrices u_{ik} and v_{ik}:

$$\epsilon_{ik} = u_{ik} + v_{ik}, \tag{5.19}$$

with the definition of u_{ik} and v_{ik} given by:

$$u_{ik} = \frac{1}{2}i\,\eta_{iklm}u^{lm} = \frac{1}{2}i\,\eta_{iklm}g^{lr}g^{ms}u_{rs}, \quad v_{ik} = -\frac{1}{2}i\,\eta_{iklm}v^{lm} \tag{5.20}$$

$$\text{and} \quad u_{ik} = v_{ik}^*.$$

The η_{iklm} is the completely antisymmetric Levi-Civita symbol. They next introduced matrices b and c:

$$b_{ik} = g_{ik} + u_{ik}, \qquad c_{ik} = g_{ik} + v_{ik}. \tag{5.21}$$

It followed from (5.20) that $b_{ik} = c_{ik}^*$. It is important to note that the indices of b_{ik} and c_{ik} were raised and lowered by the spacetime metric g_{ik}.

From (5.19) and (5.21) Einstein and Mayer concluded that any infinitesimal Lorentz transformation a of a vector can be decomposed into two matrices b and c of the form (5.21):

$$a = bc. \tag{5.22}$$

They further found that this decomposition is unique, that is, for each a there is a unique b and a unique c into which a is decomposed. It turned out that b and c commute:

$$bc = cb. \tag{5.23}$$

Einstein and Mayer continued to show that the decomposition (5.22) can be extended to finite Lorentz transformations. To understand this result, we again turn to the definitions (5.20) and (5.21). According to Einstein and Mayer, all u_{ik} can be obtained from six fundamental four-by-four matrices: in our notation these are the same as the N_n defined by (5.12), supplemented by the set iN_n. We have seen earlier that these six operators satisfy the Lie algebra of the Lorentz group. They therefore generate a group $\mathcal{B}$ with elements b, which is isomorphic to the Lorentz group. Similarly, $v_{ik} = u_{ik}^*$ can be obtained from N_n^* and $-iN_n^*$, and these together also give the Lie algebra of the Lorentz group. They generate a group $\mathcal{C}$ with elements c, which is again isomorphic to the Lorentz group.

We can now easily illustrate that b times c gives a Lorentz transformation a on a four-vector (the parameters α_n are complex numbers, $\alpha_n = \alpha_{nR} + i\alpha_{nI}$);

$$a = b \cdot c = \exp\left(\sum_n \alpha_n^*(iN_n)\right)\exp\left(\sum_n \alpha_n(-iN_n^*)\right)$$

$$= \exp\left(\sum_n (\alpha_{nR}(iJ_n) + \alpha_{nI}(iK_n))\right). \tag{5.24}$$

The J_i and K_i were chosen in a four-by-four imaginary representation, so a is a real valued four-by-four matrix giving a Lorentz transformation on a vector. Since N_n and N_n^* commute, it follows that b and c commute. We see also that the matrices b and c are uniquely associated with a particular element of the Lorentz group.[44]

With $\mathcal{B}$ and $\mathcal{C}$, Einstein and Mayer had found new ways to represent the Lorentz group, and this they used to introduce the semivector. We have just seen that a vector λ_k transforms as

$$\lambda'^k = a_i^k \lambda^i. \tag{5.25}$$

With the isomorphy $a_{ik} \to b_{ik}$ we can now define a contravariant *semivector of the first kind* ρ^s as the object that transforms under Lorentz transformations

---

[44] Note however, if $a = bc$ then also $a = (-b)(-c)$. Therefore, if the groups $\mathcal{B}$ and $\mathcal{C}$ are to be truly isomorphic to the Lorentz group one needs to identify b and $-b$, and c and $-c$. This is a consequence of the fact that the Lorentz group is equivalent to $SL(2,\mathbb{C})/\mathbb{Z}_2$ instead of $SL(2,\mathbb{C})$.

according to:[45]

$$\rho'^s = b_t^s \rho^t. \tag{5.26}$$

Analogously, the contravariant *semivector of the second kind* σ^s transforms in the conjugate $\mathcal{C}$ representation:

$$\sigma'^s = c_t^s \sigma^t. \tag{5.27}$$

So the complex conjugate of a contravariant semivector of the first kind is a contravariant semivector of the second kind, and vice versa.

These semivectors were Einstein's alternative to the spinor. They have four components that are raised and lowered with the familiar spacetime metric g_{ik}. In this respect, one could think of the semivector components as lying in spacetime. Since Weyl spinors have just two internal indices, one might consider it problematic to give them an "intuitive" or "realist" interpretation; possibly, Einstein therefore thought the semivector to be a mathematically more natural concept.

5.2.3 Unification of electrons and protons

In their second semivector paper, the first paper that appeared in Amsterdam, Einstein and Mayer were on the lookout for the most general Dirac system possible for the semivector. They found:

$$\begin{cases} E^{rst}(\psi_{s,r} - ie\psi_s A_r) = \bar{C}^{tk}\chi_k \\ E^{rst}(\chi_{t,r} - ie\chi_t A_r) = -C^{ks}\psi_k. \end{cases} \tag{5.28}$$

The system contained two semivectors, ψ_s and χ_t, where ψ_s is a semivector of the first kind and χ_t is a semivector of the second kind. The elements of E were just dimensionless numbers. The mass matrix C will play an important role in what follows.

The most general mass term in the Lagrangian which is compatible with the symmetries of the theory is of the form:

$$C_{sk}\psi^s\bar{\chi}^k - \bar{C}_{sk}\bar{\psi}^s\chi^k. \tag{5.29}$$

For the term to be a Lorentz-scalar, the mass matrix C has to be invariant under b transformations of its indices.[46]

---

[45] Equation (5.24) implies that a full rotation of a vector around the z-axis is given by $\alpha_{3R} = 2\pi$, all other angles zero. The associated b is: $b = \exp i\pi (J_3 + iK_3)$. This transforms the semivector not only with an overall minus sign (as for the spinor), but also a boost.

[46] That is, $C \to Cbb = C$, see Einstein and Mayer (1933a), p. 499.

From this demand it followed that C cannot be anything but a c-matrix. The general form of c-matrices in $\mathcal{C}$ was given by $\tilde{C}g_{sk} + v_{sk}$, with $\tilde{C}$ complex valued and v_{sk} as in (5.20). So Einstein and Mayer found for C:

$$C_{sk} = \tilde{C}g_{sk} + v_{sk}. \tag{5.30}$$

But in no way had it been fixed which c-matrix they were dealing with, and as the c-matrices form a group, the mass term (5.29) remains a Lorentz scalar if C is multiplied by a c-transformation. So the semivector Lagrangian was equivalent to a Lagrangian with

$$C'^{mt} = C^{mk}c_k^t. \tag{5.31}$$

Because of the symmetries in the definition of v, i.e. (5.20), C was fixed by four complex constants C_{11}, C_{12}, C_{13}, C_{14}. Einstein and Mayer exploited the property that C is defined up to c-transformations to reduce the number of constants in the general form (5.30). In this way they arrived at

$$C = \begin{pmatrix} ia & 0 & 0 & -ib \\ 0 & ia & b & 0 \\ 0 & -b & ia & 0 \\ ib & 0 & 0 & -ia \end{pmatrix}. \tag{5.32}$$

Now there were just three real constants left in the theory: the elementary charge e, and the constants a and b. So from the most general Dirac equation with four complex constants in C, Einstein and Mayer had arrived at the simplest possible equivalent equation that has just three real constants.[47]

The reduction procedure is equivalent to applying c-transformations on the indices of ψ^k, the semivector of the first kind[48] (so it is not a Lorentz transformation on the semivector, since then it would have to transform with b):

$$\psi_k' = c_k^t \psi_t. \tag{5.33}$$

This will prove to be an important observation when we later discuss the reception of Einstein's results.

The third semivector paper also came out in Amsterdam. The contents of this paper were presented, presumably by Ehrenfest, at the Amsterdam Academy's meeting of 24 June 1933, just two weeks after Einstein's lecture in Oxford. This paper contained the results that Einstein discussed in his Herbert Spencer lecture

---

[47] See Einstein and Mayer (1933a), p. 515.
[48] A similar reduction of the Einstein–Mayer reduction of the semivector Dirac equation can be found in Ullmo (1934).

(Einstein and Mayer had these results at least two weeks before the lecture; Einstein mentioned them in letters to Mayer, sent from Oxford and dated 3 and 4 June).[49]

Einstein and Mayer split the Dirac system for semivectors (5.28) into two Dirac systems for spinors. They did so by using a v-matrix, given by;

$$\tilde{v} = \begin{pmatrix} 0 & 0 & 0 & -i \\ 0 & 0 & 1 & 0 \\ 0 & -1 & 0 & 0 \\ i & 0 & 0 & 0 \end{pmatrix}. \tag{5.34}$$

Note that $\tilde{v}_{ts}\tilde{v}_k^s = -g_{tk}$ and that $\tilde{v}$ has the same off-diagonal components as (5.32), up to the factor b. With $\tilde{v}$ one could decompose a semivector ψ_s of the first kind into two independent parts:

$$\psi_s = \frac{1}{2}(\psi_s - i\tilde{v}_s^n\psi_n) + \frac{1}{2}(\psi_s + i\tilde{v}_s^n\psi_n) = \psi_s(\alpha) + \psi_s(\beta); \tag{5.35}$$

i.e. it decomposed the semivector into an *α-semivector* $\psi_s(\alpha)$ and a *β-semivector* $\psi_s(\beta)$. The α and β-semivector of the first kind are eigenvectors of $\tilde{v}$ with eigenvalues $+i$ and $-i$ respectively. As b and $\tilde{v}$ commute, an α- or β-semivector of the first kind remains an α-, respectively, β-semivector of the first kind under Lorentz transformations, a property that we will return to later on.

When we write out in components the α-semivector $\psi_s(\alpha)$ and β-semivector $\psi_s(\beta)$ we get, using their properties as eigenvectors of $\tilde{v}$:

$$\begin{array}{ll} \psi_1(\alpha) = -\psi_4(\alpha) & \psi_1(\beta) = \psi_4(\beta) \\ \psi_3(\alpha) = i\psi_2(\alpha) & \psi_3(\beta) = -i\psi_2(\beta). \end{array} \tag{5.36}$$

The α and β-semivectors of the second kind were defined by the complex conjugate equations: with each α-semivector of the first kind was associated through complex conjugation an α-semivector of the second kind. Since the α and β-semivectors have just two independent components, Einstein and Mayer identified them with spinors. They nevertheless preferred the semivector over the spinor, "because of its simpler transformation laws."[50]

In any case, using the decomposition of the semivector into α and β-semivectors, Einstein and Mayer could split the Dirac system (5.28) into two Dirac spinor systems. As the α and β-semivectors by (5.35) are eigenvectors of the $\tilde{v}$-matrix, and the matrix C in (5.32) was of the form $C_{ks} = iag_{ks} + b\tilde{v}_{ks}$, the α and β-semivectors were eigenvectors of the C-matrix too. This gave (note that the properties of E entailed that it changes the character of an α or β-semivector in, respectively, a β

[49] 3 June 1933, Einstein to W. Mayer, EA 18-175; 4 June 1933, Einstein to W. Mayer, EA 18-176. From these letters we learn that this result was due to Mayer.

[50] "Man sieht, dass sich die Theorie der Spinoren aus der Theorie der Semi-Vektoren ergibt. Es scheint aber, dass infolge seines einfacheren Transformationsgesetzes der Semi-Vektor dem Spinor vorzuziehen ist." As in Einstein and Mayer (1932b), p. 550.

or α-semivector):

$$\begin{cases} E^{rst}(\psi_{s,r}(\alpha) - ie\psi_s(\alpha)A_r) = -i(a-b)\chi^t(\beta) \\ E^{rst}(\chi_{t,r}(\beta) - ie\chi_t(\beta)A_r) = i(a-b)\psi^s(\alpha) \end{cases} \tag{5.37}$$

and

$$\begin{cases} E^{rst}(\psi_{s,r}(\beta) - ie\psi_s(\beta)A_r) = -i(a+b)\chi^t(\alpha) \\ E^{rst}(\chi_{t,r}(\alpha) - ie\chi_t(\alpha)A_r) = i(a+b)\psi^s(\beta). \end{cases} \tag{5.38}$$

Each set of spinor Dirac equations gave the equation of motion of a wave field: the first set of equations suggested that there were de Broglie wave solutions with rest mass

$$m_1^2 = (a-b)^2, \tag{5.39}$$

and for the second set

$$m_2^2 = (a+b)^2. \tag{5.40}$$

This was the result that Einstein was so pleased with in Oxford. The first set could be interpreted as the Dirac equation for the electron, and the second, upon complex conjugation, could be interpreted as the Dirac equation for the proton; that is, the original negative mass solution was a positive mass solution of the conjugate equation, in which the sign of e had changed. In the first Amsterdam paper – that did allude to the possibility of two solutions with differing masses but not yet had the split of (5.28) into two fundamental spinor systems – Einstein and Mayer showed themselves highly satisfied with these results. Not only had they found "for the first time an explanation for the fact that [...] there are two electrical elementary particles of different mass, the electrical charges of which have an opposite sign," they also ventured that there may be not only a positron, but even something like an anti-proton: "The fact that apart from positive values there are also negative values for the mass constant m may be related to the apparent occurrence of 'positive electrons'. These in any case could be seen as electronegative particles with negative ponderable mass. According to the theory, the same might also be expected for the protons."[51]

Understanding why Einstein believed that an explanation had been found for the occurrence of two fundamental masses becomes possible when we turn to his contemporary methodological ideas. Firstly, he had found an alternative for the spinor that was more mathematically natural. Its simplest – or most general

---

[51] "[...] zum ersten Male eine Erklärung dafür [...], dass es zwei electrische Elementarteilchen verschiedener Masse gibt, deren elektrische Ladungen entgegengesetztes Vorzeichen besitzen." [...] "Dass neben den positiven auch die negativen m Werte als Massenkonstante auftreten, hängt vielleicht mit dem scheinbaren Auftreten "positiver Elektronen" zusammen, die allerdings als elektronegative Teilchen von negativer ponderabler Masse anzusehen wären. Entsprechendes wäre nach der Theorie auch für die Protonen zu erwarten." As in Einstein and Mayer (1933a), p. 515.

Lorentz-invariant – Lagrangian and equation of motion contained a mass matrix C which, as if by a miracle, was defined up to c-transformations. By using this symmetry, the equation could be further simplified and be put in a form in which just three constants remained. They finally arrived at two field equations for two particles with a different mass, but with charges of the same magnitude. That meant that looking for the most natural alternative for the spinor, and this alternative's simplest equation, had produced a unified description of the electron and proton. It had even prompted an empirical prediction: from the symmetry properties of the C-matrix followed, by diligent deduction, Einstein's prediction that there had to be, not only a positron, but also an anti-proton.

The split into two fundamental spinors in fact toned down Einstein and Mayer's enthusiasm for the semivector somewhat, since its simplest formulation brought one back to the spinor. They now realized that semivector calculus was actually not necessarily simpler than spinor calculus: nevertheless, "this analysis does not imply [...] that the semivector theory has now proven to be unnecessary." Namely, Einstein and Mayer still believed the unified description of the electron and proton to be a major achievement of the semivector formalism: "From the standpoint of the spinor theory it cannot be understood why in nature there are precisely *two* different elementary inertial masses with apart from the sign an equal electrical charge."[52] Since to Einstein unification and explanation amounted to the same thing, he felt that the understanding of the relations between electrons and protons had been truly enhanced with their unified description in the semivector Dirac system. As he wrote to Paul Ehrenfest: "The theory of semivectors leads to a completely natural explanation of the difference in mass of positive and negative electricity of equal charge."[53]

---

[52] "[D]iese Zerlegung besagt [...] nicht, dass sich nun die Semi-Vektoren-Theorie als unnötig erwiessen hätte." – "[V]om Standpunkt der Spinorentheorie [ist] nicht verständlich, warum es in der Natur gerade *zwei* verschiedene elementare Trägheitsmassen mit abgesehen vom Vorzeichen gleich grosser elektrischer Ladung gibt." As in Einstein and Mayer (1933b), p. 619; Einstein and Mayer's emphasis.

[53] "Die Theorie der Semi-Vektoren führt zu einer ganz ungezwungenen Erklärung der Verschiedenheit der Masse der positiven und negativen Elektrizität bei gleicher Elementarladung." 1 May 1933, EA 10-249. Einstein similarly wrote to Wander de Haas (7 May 1933, EA 70 493): "Scientifically I have found together with my (Walther) Mayer a very natural generalization of the Dirac formulas that makes understandable that there are two ponderable elementary masses, but only a single electrical one." ("Wissenschaftlich habe ich mit meinem (Walter) Mayer eine sehr natürliche Verallgemeinerung der Dirac'schen Formeln gefunden, welche es begreiflich macht, dass es zwei ponderable Elementarmassen, aber nur eine elektrische gibt.") Einstein's ideas about understanding and unification are further well reflected in a letter he sent from Oxford to advertise the work to Max von Laue: "With Mayer I have found a generalization of the Dirac theory that explains that there are two kinds of particles with differing masses (the ratio of the masses (numerical value) is not logically determined), the same charge magnitude and opposite charge sign. The theory is of great mathematical correctness and consistency, the paper will soon appear in the Dutch academy proceedings." ("Mit Mayer habe ich eine Verallgemeinerung der Dirac'schen Theorie gefunden, die erklärt, dass es zwei Teilchenarten mit verschiedener Masse (das Massenverhaltnis (Zahlenwert) ist logisch nicht bestimmt), gleichem Betrag der Ladung und entgegengesetztem Vorzeichen der Ladung gibt. Es ist eine Theorie von grosser mathematischer Folgerichtigkeit, die Arbeit erscheint bald in der holländischen Akademie.") Einstein to M. von Laue, 26 May 1933, EA 16-090. The remark that the mass ratio had not been fixed will actually prove to be an ominous observation; see Section 5.3.

5.2.4 Results of the semivector theory and Einstein's methodology

One of the goals of the semivector formalism had been to formulate the Dirac equation in a curved spacetime. The diffeomorphism group has no spinorial representation, and the semivector might just have provided the means to work around this in a natural way. The incorporation of the Dirac equation into general relativity was dealt with in Einstein and Mayer's first semivector paper. However, their solution looks familiar: the semivector components lay in a flat spacetime that was tangent to the curved spacetime, and the covariant derivative of the semivector was obtained by adding the analog of a spin-connection to the flat space derivative operator. This seems very similar to the procedure for spinors, outlined earlier by a number of Einstein and Mayer's contemporaries.[54] Einstein and Mayer nevertheless preferred the semivector version over the spinor equation, qualifying the latter as less natural.[55] But as the procedure for both objects was so much alike, it is not entirely obvious why the semivectors should be preferred. Perhaps Einstein's preference was due to the fact that the four semivector components could more readily be thought of as lying in a "real" four-dimensional tangent spacetime, and were therefore considered to be more "natural."[56]

In any case, Einstein and Mayer indeed succeeded in incorporating the Dirac equation into general relativity, but this is not the result that we wish to address here. The unification of the electron and proton was a much more telling example of the success of striving after mathematical simplicity and naturalness. When we look at the correspondence between Einstein and Mayer, it becomes yet more clear that this methodological conviction of Einstein was the prime inspiration for the semivector research. Of course, as the latest foremost achievement of his methodology, the unification result was to figure prominently in the Herbert Spencer lecture.

The Einstein–Mayer correspondence contains many letters that suggest that the Spencer lecture methodology motivated the work on semivectors – "This is an

---

[54] See for example Weyl (1929), Infeld and van der Waerden (1933). Einstein and Mayer referred to the work of Infeld and van der Waerden (1933) and conceded that these authors's procedure is essentially the same. For more historical references on the Dirac equation in curved spacetime, see Kichenassamy (1992).

[55] The second Amsterdam paper gives as one of the fruits of the semivector calculus "the possibility of incorporation in the construction of general relativity, which a pure spinor theory does not have so naturally" ("[die] Einbaumöglichkeit in das Gebäude der allgemeinen Relativität, die eine reine Spinoren Theorie so zwanglos nicht kennt"). As in Einstein and Mayer (1933b), p. 619.

[56] The 1934 paper on semivectors in the *Annals of Mathematics* (Einstein and Mayer, 1934) proposed a way to write the semivector Dirac equation in curved spacetime without making use of a local orthogonal tetrad, unlike spin-connection type procedures. Einstein and Mayer however said that they did not favor their newest result over the procedure they had formulated earlier.

entirely natural generalization [...]" – "This formulation of concepts is extraordinarily natural" – "[...] that the introduction of spinors is indeed unnatural."[57] There is one example that we wish to single out and study in some detail, as that occasion illustrates particularly well the role of Einstein's contemporary methodological ideas.

Einstein and Mayer were not the only ones thinking about the relation between electrons and protons, and the ratio of their masses – Sir Arthur Eddington had picked up the same problem.[58] Einstein learned of his work early in 1933 in a seminar that he attended during a visit to the California Institute of Technology.[59] Initially he wondered whether he and Mayer should try to join in with Eddington's approach. A week after the seminar, he decided against this. He wrote to Walther Mayer:

I no longer think that following Eddington's equation is the right way. [...] The *mathematically* most natural way seems to me to be the following. Analogy to the equation $R_{ik} = 0$ of the pure gravitational field. Instead of the vector curvature [presumably a reference to the Riemann tensor or one of its contractions, e.g. the Ricci scalar] one starts with the semivector curvature. [...] In any case, one can thus characterize manifolds by the curvature alone, without introducing field variables in an arbitrary way. This is really the point of a *"unified field theory,"* which represents the mathematically most natural kind of field theory. This seems to me the direction in which we should look. I have tried to study Eddington's new work on multiple electrons; I am aghast at the unnaturalness of the structures to which he needs to resort. This is not how nature sensibly can be understood, despite the apparent success in single instances.[60]

---

[57] "Dies ist eine ganz natürliche Verallgemeinerung [...]" – "Diese Begriffs-Bildung ist eine ungemein natürliche." – "[...] dass die Einführung der Spinoren tatsächlich unnatürlich ist." Respectively, Einstein to Mayer, 8 January 1933, EA 18-152; Einstein to Mayer, 20 January 1933, EA 18-154; Einstein to Mayer, 26 March 1933, EA 18-164. It is not difficult to find more examples that suggest the same influence: "If correct, then we have [...] a complete foundation for the theory of electrons that is incomparably simpler than Dirac's." ("Stimmt es, so haben wir [...] eine vollständige Grundlage für die Theorie des Elektrons, die unvergleichlich einfacher ist als diejenige von Dirac," 17 November 1932, Einstein to W. Mayer, EA 18-139). "So there is [...] no reason to consider such complicated systems of equations as Dirac did." ("Es besteht also [...] kein Grund dafür, so komplizierte Gleichungssysteme ins Auge zu fassen wie Dirac," 8 December 1932, Einstein to W. Mayer, EA 18-143.)

[58] For a historical discussion of Eddington's thought on this issue, see Chapter 9, "The proton–electron mass ratio," in Kilmister (1994).

[59] Einstein in his letter to Mayer did not reveal who was the speaker; 14 February 1933, Einstein to W. Mayer, EA 18-160.

[60] "[...] die Anlehnung an Eddingtons Gleichung scheint mir nicht mehr der rechte Weg zu sein [...]. Der *mathematisch* natürlichste Weg scheint mir folgender zu sein. Analogie zu der Gleichung $R_{ik} = 0$ des reinen Gravitationsfeldes. Statt von der Vektorkrümmung geht man von der Semi-Vektor-Krümmung aus. [...] Jedenfalls kann man so Mannigfaltigkeiten aus der Krümmung allein charakterisieren, ohne Feldvariable willkürlicher Art nebeneinander einzuführen. Dies ist doch der Sinn einer „*einheitlichen* Feldtheorie", welche den mathematisch natürlichsten Typ einer Feldtheorie repräsentiert. Dies scheint mir die Richtung, in der wir suchen müssen. Ich habe versucht, Eddingtons neue Arbeiten über mehrere Elektronen zu studieren; ich bin entsetzt über die Unnatur der Bildungen zu denen er seine Zuflucht nehmen muss. So kann die Natur nicht vernünftig erfasst werden, trotz der scheinbaren Erfolge im Einzelnen." 23 February 1933, Einstein to W. Mayer, 18-163. Einstein's emphasis.

Einstein was not alone in his dislike for Eddington's ideas – only a few years before Wolfgang Pauli had even qualified Eddington's program as "complete nonsense; more precisely, as romantic poetry, not as physics."[61] Eddington was charting out his own alternative to developments in quantum field theory; he was fascinated by possible generalizations of the Dirac algebra and in this way hoped to uncover clues about for instance the value of the fine-structure constant. He had also tried to formulate an equation that related the electron–proton mass ratio to the curvature of the universe and its total number of protons.[62]

In any case, the essential issue to observe in the excerpt above is that when Einstein was looking for direction, he concluded that he should aim for the "mathematically most natural" approach. He echoed his lecture at Oxford – four months before the occasion however, and with the intention of finding the direction his research should take. Yet this letter uncovers still more about Einstein's rationale, as he continued with a revealing remark:

One should look for the mathematically most natural structures, without initially being bothered too much about the physical, as this brought the desired result in gravitation theory.[63]

Indeed, Einstein grounded his epistemological conviction in his re-shaped recollection of the discovery of the general theory of relativity. We see that that profound experience, the exhilarating November month and the preceding deep struggles with the Entwurf theory, here produced a direct influence on his research. When he wrote to Mayer that they should look for the equivalent of the $R_{ik} = 0$ equation from general relativity, he was convinced that this would produce the mathematically most natural semivector equation.[64] It can be no surprise that he wanted the semivector equivalent of the Ricci or Riemann tensor; these had eventually delivered him general relativity, according to his remolded recollections. One had to follow the same path: as Einstein wrote Mayer, an "unnatural" set of laws may be successful in some instances, but will surely fail if one wants a comprehensive understanding of nature.

---

[61] "[D]ie Eddingtonsche [...] Arbeit halte ich jetzt für kompletten Unsinn; genauer gesagt: für romantische Poesie und nicht für Physik." ("I now regard the Eddington [...] work as complete nonsense; more precisely, as romantic poetry, not as physics.") W. Pauli to Oskar Klein, 18 February 1929, Doc. 216, pp. 488–492 in Hermann *et al.* (1979), on p. 491; p. 116 in Kilmister (1994). According to the Eddington scholar Clive W. Kilmister, Pauli's position differed only in intemperate statement from the majority view.

[62] For more on these projects, see Eddington (1933).

[63] "Man muss die natürlichsten mathematischen Bildungen aufsuchen, ohne sich zunächst um das Physikalische zu bekümmern, wie es bei der Gravitationstheorie zum Ziel geführt hat." 23 February 1933, Einstein to W. Mayer, 18-163.

[64] In fact, the above remarks may very well have quite literally shaped what was to be presented in the first Amsterdam paper: the "most general" Dirac equation compatible with the symmetries of the theory (5.28) was derived from the variational principle, applied to the most general Lagrangian possible for the semivector; see Einstein and Mayer (1933a). However, they did not lead immediately to the introduction of the C-matrix: in this letter Einstein contemplated the possibility of making the E-constants into spacetime dependent fields.

In his private letter Einstein ventured a bit further than he might have dared to in his public address at Oxford: he confided to Mayer not that experience is the alpha and omega of all knowledge, but rather that one should not be bothered too much about the "physical" – not immediately in any case – because this was how he had arrived at general relativity. We have seen that Einstein found the field equations through an intricate interplay of mathematical and physical demands, but that, according to his later recollections, the process was only finalized when the mathematical requirement of general covariance prevailed. We have further argued that following his emotional conquest, he gradually rearranged his epistemology and accompanying methodology in line with those recollections, just as in turn the recollections and his methodological stances were being molded by his current research practice. The semivector theory is both an example of what his re-shaped methodology was to produce, and the kind of theorizing it was a reflection of. The fundamental laws of the semivector theory were not found by any induction from the phenomena; they were seemingly invented in a way that was not constrained by experience at all, but appear to be entirely "free inventions of the human intellect" – free inventions that were guided by the conviction that nature satisfies the simplest imaginable mathematical laws.

If we look at the Solovine schema, then what should follow after the formulation of the simplest possible semivector equation is solid logical deduction. It is by deduction, towards ever greater mathematical simplicity (i.e. the reduction of the number of constants in the Dirac equation), that Einstein and Mayer found their major result: the unified description of the electron and proton. Furthermore, by deduction – we learn from their correspondence – they wanted to attain an even bigger prize: the quantization of charge. Einstein would then finally be on his way to a classical field theory that could undercut quantum mechanics.

However, Einstein and Mayer never managed to derive solutions with a discrete charge spectrum from the semivector equations. In principle, the idea was to find a non-singular solution of the Dirac equation, combined with the Maxwell equations. Einstein believed that such a non-singular solution would open numerous possibilities:

If we were to succeed in integrating the equations without singularities, then also the numerical relation between e^2/c and h should follow. [...] Ceterum censeo: I believe in the field theory and hold the statistical theory to only be an intermediate stage.[65]

Now the angular momentum of the electron (the "spin") also needs to be computed. [...] Only if we have all this can we go look for an integral for the total charge. If that integral

---

[65] "Wenn es gelünge, die Gleichungen singularitätsfrei zu integrieren, so müsste sich auch die numerische Beziehung zwischen e^2/c und h ergeben. [...] Ceterum censeo: Ich glaube an die Feldtheorie und halte die statistische nur für ein Übergangsstadium." Einstein to W. Mayer, 21 December 1932, EA 18-149.

can be proven to be integer valued, then the quantum riddle is essentially solved, notably by a true field theory.[66]

But in the course of 1933 Einstein's frustration grew as the non-singular solution proved elusive. At the same time, he found the result of ever greater importance:

With regard to the interpretation of the integer-valuedness of the electrical charge, I have not made a single step forward. It has to be possible to somehow deduce an equation, by elimination from the Dirac equations, that no longer contains differentiations with respect to x^4. Only from such an equation something can be deduced that applies to just one slice $x^4 = \text{const[ant]}.$[67]

Two days later, on the fourth of June, he wrote:

It annoys me that our Dirac equations do not contain a condition for a time slice. That is, it looks as if J_4 can be chosen arbitrarily on a slice of $x^4 = \text{const}$, that is, as if $\int J_4 dx_1 dx_2 dx_3$ can be given any value, in contradiction to the atomicity of electricity. From the physical point of view, this is in my opinion the key problem![68]

A week later:

The question is: does the integer-valuedness of $\int J_4 dx_1 dx_2 dx_3$ follow from our equations for solutions with at ∞ vanishing, singularity free $\psi \chi$? [...] If this question is not answered affirmatively, then the theory in its current form cannot be taken as a model of reality.[69]

The attempt to derive the discrete charge spectrum was unsuccessful and was not reported on in any article (although the general idea was outlined in the Oxford lecture, Einstein on that occasion ventured that the desired non-singular particle solution might even provide a means to understand the Heisenberg uncertainty relations in the context of classical field theory; we will briefly discuss what he said in Chapter 7). The last three letters above were actually written in Oxford, the last letter in fact exactly one day after he had given the Herbert Spencer lecture. The lecture suggests that Einstein's confidence in the semivector theory was at the time still quite firm; yet his correspondence suggests that that mood was already waning.

---

[66] "Nun muss [...] auch noch das Impulsmoment des Elektrons (der „Drall") ausgerechnet werden. [...] Erst wenn wir dies alles haben, können wir an die Suche nach einem Integral für die Gesamtladung gehen. Ist dieses als ganzzahlig zu beweisen, dann ist das Quantenrätsel in der Hauptsache gelöst, und zwar durch eine richtige Feld-Theorie." Einstein to W. Mayer, 31 May 1933, EA 18-172.

[67] "[B]ezüglich der Deutung der Ganzzahligkeit der elektrischen Ladung bin ich keinen Schritt weiter gekommen. Es müsste doch aus den Dirac-Gleichungen durch Elimination irgendwie eine Gleichung abzuleiten sein, welche keine Differentiationen nach x^4 mehr enthält. Nur aus einer solchen Gleichung kann etwas geschlossen werden, das nur einen Schnitt $x^4 = \text{konst.}$ betrifft." 2 June 1933, Einstein to W. Mayer, EA 18-173.

[68] "Es irritiert mich aber, dass unsere Dirac-Gleichungen kein Bedingung für den zeitlichen Schnitt zu enthalten scheinen. D.h. es sieht so aus, wie wenn man in einem Schnitt $x^4 = \text{konst.}$, J_4, also auch $\int J_4 dx_1 dx_2 dx_3$ nach Belieben wählen könnte, was der Atomistik der Elektrizität widerspräche. Dies ist nach meiner Ansicht vom physikalischen Gesichtspunkte aus betrachtet das Kernproblem!" 4 June 1933, Einstein to W. Mayer, EA 18-176.

[69] "Die Frage ist: folgt die Ganzzahligkeit von $\int J_4 dx_1 dx_2 dx_3$ aus unseren Gleichungen für Lösungen mit im ∞ verschwindenden singularitätsfreien $\psi \chi$? [...] Wenn diese Frage nicht positiv sich beantwortet, kann die Theorie in ihrer gegenwärtigen Form nicht als Modell der Wirklichkeit angesehen werden." 11 June 1933, Einstein to W. Mayer, EA 18-177.

5.3 Reception of the semivector

During his stay on the European continent early in 1933, Einstein traveled to Switzerland to see his mentally ill son Eduard. It is quite possible that on this occasion he also visited Wolfgang Pauli in Zurich, and that they discussed their latest ideas. Pauli had taken an interest in the mathematical issues arising when one tries to incorporate the Dirac equation in general relativity,[70] and he would likely have wanted to stay informed about Einstein's current work. Sometime in June Einstein sent his latest semivector papers to Pauli. On 16 July, Pauli replied:

Dear Herr Einstein!

Thank you very much for sending me the galley proofs of your new work with Mayer. I have now studied it carefully, yet I have very serious physical reservations regarding your basic idea.[71]

Pauli pointed to unphysical interference terms that could occur in the expression for the charge if one adds up two fundamental particles in one semivector. The second Amsterdam paper, however, showed that the split of the semivector into two spinor systems is such that these interference terms do not arise. He had another objection, however, and this in fact proved to announce the demise of the semivector:

From a formal point of view, I do not think [...] I will greatly befriend your semivectors, because they do not correspond to an *irreducible* representation of the rotation [i.e. Lorentz] group. And at the moment I am having Herr Bargmann rewrite the equations of your latest work in spinors.[72]

Valentin Bargmann was at the time a doctoral student in Zurich and under the guidance of Pauli, he made a critical assessment of the semivector (more on Bargmann can be found in Chapter 6). Bargmann rewrote all of Einstein and Mayer's semivector work in van der Waerden's notation. His article was submitted to the journal *Helvetica Physica Acta* less than four months after Pauli's letter and appeared in print early in 1934.[73]

Basically, Bargmann followed through on the realization that a semivector contains two spinors,[74] in other words, on Pauli's observation that the semivector representation was not an irreducible representation of the Lorentz group. He

---

[70] Later that year Pauli published work that studied this problem in a five-dimensional spacetime (Pauli, 1933b).

[71] "Lieber Herr Einstein! Ich danke Ihnen noch sehr für die Zusendung der Korrekturen Ihrer neuen Arbeit mit Mayer. Ich habe sie nun sorgfältig studiert, habe aber die grössten physikalischen Bedenken gegen Ihre Grundidee." W. Pauli to Einstein, 16 July 1933, EA 19-173, Doc. 315, pp. 189–190 in von Meyenn (1985).

[72] "Mit den Semivektoren kann ich mich [...] auch formal nicht sehr befreunden, da sie keiner *irreduziblen* Darstellung der Drehgruppe entsprechen. Und ich lasse jetzt die Gleichungen Ihrer neuen Arbeit von Herrn Bargmann in Spinoren umschreiben." W. Pauli to Einstein, 16 July 1933, EA 19-173; Doc. 315, pp. 189–190 in von Meyenn (1985), emphasis as in original.

[73] See Bargmann (1934).

[74] Related work had been done earlier by J. A. Schouten (1933). Schouten's article did not however address the reduction of the Dirac equation. We thank Hubert Goenner for pointing out Schouten's work to us.

started by showing that flat space, $\mathbb{R}^4$, can be divided into two two-dimensional subspaces $M_{\dot{1}}$ and $M_{\dot{2}}$, where both $M_{\dot{\lambda}}$ are invariant under the transformation b_k^i, the matrix used to perform a Lorentz transformation on a semivector of the first kind. This means that any vector z^l can be split by using projection operators $S_{\dot{1}}$ and $S_{\dot{2}}$:

$$z^l = S_{\dot{1}m}^l z^m + S_{\dot{2}m}^l z^m \qquad (5.41)$$

where $S_{\dot{1}m}^l z^m \in M_{\dot{1}}$, $S_{\dot{2}m}^l z^m \in M_{\dot{2}}$ and $S_{\dot{\lambda}m}^l b_k^m = b_m^l S_{\dot{\lambda}k}^m$. In precisely the same way, $\mathbb{R}^4$ can be split into two spaces N_μ by operators T_μ that commute with the transformation c (i.e. a Lorentz transformation for a semivector of the second kind). The split by S is the same as the split of semivectors of the first kind into α and β-semivectors of the first kind. Likewise, the split by T is the same as the split of semivectors of the second kind into α and β-semivectors of the second kind.

Bargmann then argued that any vector z^l can be decomposed on four basis vectors, two of which span $M_{\dot{1}}$ (i.e. $e_{\dot{1}1}^l$ and $e_{\dot{1}2}^l$) and two of which span $M_{\dot{2}}$ (i.e. $e_{\dot{2}1}^l$ and $e_{\dot{2}2}^l$). In different combinations, the same set of four vectors span N_1 (i.e. $e_{\dot{1}1}^l$ and $e_{\dot{2}1}^l$) and N_2 (i.e. $e_{\dot{1}2}^l$ and $e_{\dot{2}2}^l$). The decomposition of the vector is given by

$$z^l = \sum_{\dot{\lambda}\mu} S_{\dot{\lambda}n}^l T_{\mu m}^n z^m = (\xi^{\dot{\lambda}\mu}) e_{\dot{\lambda}\mu}^l. \qquad (5.42)$$

One can in fact take as basis vectors (e^l) the Pauli matrices (σ^l). This means (as also follows from (5.7) by using $\sigma_{\dot{\lambda}\mu}^l \sigma_m^{\dot{\lambda}\mu} = \delta_m^l$):

$$z^l = \xi^{\dot{\lambda}\mu} \sigma_{\dot{\lambda}\mu}^l. \qquad (5.43)$$

A semivector of the first kind u^m is decomposed in exactly the same way as the vector z^l. According to its definition, it transforms with b_m^l under a Lorentz transformation. An α or β-semivector of the first kind stays an α or β-semivector of the first kind under b: since $S_{\dot{\lambda}} b = b S_{\dot{\lambda}}$, the α-semivector $\sigma_{\dot{1}1}^l$ transforms into a linear combination of $\sigma_{\dot{1}1}^l$ and $\sigma_{\dot{1}2}^l$, both in $M_{\dot{1}}$. The same applies to the α-semivector $\sigma_{\dot{1}2}^l$: it also transforms into a linear combination of $\sigma_{\dot{1}1}^l$ and $\sigma_{\dot{1}2}^l$. In other words:

$$b_m^l \sigma_{\dot{1}\nu}^m = \beta_\nu^\mu \, \sigma_{\dot{1}\mu}^l. \qquad (5.44)$$

This means that

$$b_m^l \sigma_{\dot{1}\nu}^m \xi^{\dot{1}\nu} = (\beta_\nu^\mu \xi^{\dot{1}\nu}) \sigma_{\dot{1}\mu}^l. \qquad (5.45)$$

The same argument applies to β-semivectors of the first kind, i.e. $\sigma_{\dot{2}\mu}^l$, and Bargmann argued that these transform with the same matrix β_ν^μ. So we can write the Lorentz transformation of u^m as:

$$b^l_m u^m = (\beta^\mu_\nu \xi^{\dot\lambda\nu}) \sigma^l_{\dot\lambda\mu}. \tag{5.46}$$

By following an analogous line of reasoning Bargmann showed that the semivector of the second kind ($v^m = \rho^{\dot\lambda\mu} \sigma^m_{\dot\lambda\mu}$) transforms as:

$$c^l_m v^m = (\gamma^{\dot\lambda}_{\dot\rho} \rho^{\dot\rho\mu}) \sigma^l_{\dot\lambda\mu}. \tag{5.47}$$

Thus, in the case of the semivector of the second kind, the dotted index is rotated under a Lorentz transformation.

As we know, the Lorentz transformation on a vector z^l can be decomposed into two transformations: $a^l_m = b^l_n c^n_m$. Since S commutes with b and T commutes with c, the same argument as above tells us that when applying b, one can keep the dotted index of the vector fixed, and when applying c one can keep its un-dotted index fixed. The vector thus transforms as:

$$a^l_m z^m = (\beta^\alpha_\nu \xi^{\nu\dot\lambda} \gamma^{\dot\alpha}_{\dot\lambda}) \sigma^l_{\alpha\dot\alpha}, \tag{5.48}$$

with, as follows from $c_{ik} = b^*_{ik}$, $\gamma^{\dot\alpha}_{\dot\lambda} = \beta^{*\dot\alpha}_{\dot\lambda}$. Since a^l_m is a Lorentz transformation, it must preserve the norm of z^l ($z'^l z'_l = z^l z_l$). This means that $\det(\xi'_{\dot\alpha\alpha}) = \det(\xi_{\dot\alpha\alpha})$, from which follows $\beta, \gamma \in SL(2,\mathbb{C})$.

Indeed, by applying $\beta, \gamma \in SL(2,\mathbb{C})$ the vector undergoes a Lorentz transformation: this is precisely the $(\frac{1}{2}, \frac{1}{2})$-representation of the Lorentz group. This observation gives more insight into what kind of an object the semivector is. From the above it is clear that the semivector of the first kind u^l can be identified with a matrix $\xi_{\dot\alpha\alpha}$ (just as was done earlier for the vector in (5.7)); the semivector of the second kind v^l can be written as a matrix $\rho_{\dot\alpha\alpha}$. Bargmann pointed out that under Lorentz transformations these matrices go over into

$$\xi'^{\dot\alpha\alpha} = \beta^\alpha_\mu \xi^{\dot\alpha\mu} \tag{5.49}$$

$$\rho'^{\dot\alpha\alpha} = \gamma^{\dot\alpha}_{\dot\mu} \rho^{\dot\mu\alpha}. \tag{5.50}$$

In both cases one index transforms as a spinor index, that is, with an $SL(2,\mathbb{C})$ matrix. Also, in both cases one index is inert, it does not transform. Thus, the semivector is equivalent to two Weyl spinors, since $\alpha, \dot\alpha = 1, 2$. The semivector of the first kind is equivalent to two left-handed Weyl spinors (namely $\xi^{1\alpha}$ and $\xi^{2\alpha}$) and the semivector of the second kind is equivalent to two right-handed Weyl spinors (namely $\xi^{\dot\alpha 1}$ and $\xi^{\dot\alpha 2}$).

Now we can also see what the semivector representation of the Lorentz group means formally. If we think of the semivectors as vector objects with four components (i.e. as vectors in the $(\frac{1}{2}, \frac{1}{2})$ spinor-representation), we should write their transformations under the Lorentz group as follows:

$$\xi'^{\dot{\alpha}\alpha} = (1)^{\dot{\alpha}}_{\dot{\nu}} \, \xi^{\dot{\nu}\mu} \, \beta^{\alpha}_{\mu} \tag{5.51}$$

$$\rho'^{\dot{\alpha}\alpha} = \gamma^{\dot{\alpha}}_{\dot{\nu}} \, \rho^{\dot{\nu}\mu} \, (1)^{\alpha}_{\mu}. \tag{5.52}$$

The extra index transforms as if its transformation matrix is confined to the unit matrix, that is, as if it necessarily transforms under the trivial element of the Lorentz group. Einstein and Mayer had de facto introduced an extra spinor index and then constrained it to remain inert under Lorentz transformations. This aspect, however, speaks against the mathematical naturalness of the semivector, as it is not really "natural" or simple first to introduce one extra index and then to confine its transformation properties. Élie Cartan, in his famous 1937 lectures on spinors, may have been making the same point when he qualified the semivector as lacking a "geometrical definition." Among Cartan's most important work is the geometrical interpretation of groups and it is quite possible that he meant to express that there is no natural way of interpreting geometrically the inert semivector index. This could then be taken as his way of saying that the semivector was in fact a mathematically *unnatural* object.[75]

Bargmann's discussion continued with the Dirac equation. As we saw, Einstein and Mayer first produced the most general Dirac equation for semivectors, which they then rewrote by using c-matrices. In this way they reduced the number of constants in the theory to just three. Subsequently, upon splitting the semivector into α and β-semivectors, or spinors, the Dirac equation split in two as well, one equation describing the lighter electron and the other describing the heavier (anti-)proton. Bargmann, however, first wrote the most general Dirac equation for semivectors as the most general Dirac equation for spinors, by making identifications for the semivector along the lines of (5.43). He then used the γ-transformations to reduce the number of constants in the equation. So he introduced spinors before he reduced the number of constants, unlike Einstein and Mayer, who did things the other way around. Although it seemed as if the reduction of constants was a positive attribute and perhaps unique quality of the semivector description, the same general procedure now turned out to apply to spinors as well. Einstein had replaced the

[75] Cartan's 1937 lectures in Paris have been published and translated into English, and the translations have run into several editions. Cartan was the first to describe spinors, in 1913, long before their use in physics was realized. In the publication of his Paris lectures there is only one remark on the Einstein–Mayer semivector: in a footnote on p. 130 of the 1966 edition of *The Theory of Spinors* (Cartan, 1966), one finds: "A semivector of the first (second) type of Einstein and Mayer [...] has no purely geometric definition." In a talk at Caltech, early in 1933, reportedly after a question by R. C. Tolman, Einstein admitted that he could not give a geometrical or physical picture for the semivector; see Langer (1933). For completeness, we should point out that at least two researchers who studied the relation between quaternions and semivectors were in fact quite positive regarding Einstein's introduction of the semivector; see Scherrer (1934), Blaton (1935).

spinor, deemed unnatural, by the semivector, and consequently found that its most general equation of motion described rather miraculously at the same time the proton and electron – but the semivector analysis turned out to have been redundant and might as well have been carried out for spinors.

The reduction by Bargmann went through the same motions as Einstein and Mayer's: where Einstein and Mayer started by using c-transformations to bring the mass matrix C in the form (5.32), Bargmann used a $\gamma^{\dot{\alpha}}_{\beta}$ transformation on its spinor equivalent. He also observed that this procedure is the same as applying a c-transformation on the semivector of the first kind, ψ^l. He reinterpreted that transformation in the spinor notation of the semivector:

$$\psi'^k = c^k_l \psi^l \quad \rightarrow \quad \xi'^{\dot{\alpha}\alpha} = \gamma^{\dot{\alpha}}_{\dot{\beta}} \xi^{\dot{\beta}\alpha}. \tag{5.53}$$

Under a Lorentz transformation the matrix $\xi^{\dot{\alpha}\alpha}$ associated with a semivector of the first kind has its α index rotated, not its $\dot{\alpha}$ index, see relation (5.49). But in (5.53) its $\dot{\alpha}$ index is acted upon. This just means that a different superposition of the two spinors ξ^1 and ξ^2 is written in the semivector. The reduction procedure of Einstein and Mayer was thus nothing else than choosing a different superposition of the spinors in the semivector.[76]

This implied that the "most general"[77] Dirac equation for the semivector was just a very elaborate but essentially equivalent way of writing two independent spinor Dirac equations. Each of these Dirac equations has a free mass parameter and these can be put at any value. Together with the electric charge, they make up the three constants in the theory – that is, in the two separate equations. So the semivector had actually not achieved any enhanced, unified description of the electron and proton, but in effect was no more than a, not quite transparent, mix-up of the equations for two separate particles (see also Figure 5.3).[78]

It thus seems that the striving for mathematical naturalness had led Einstein astray; all that had been achieved was a tangling up of two Dirac equations, and when he and Mayer succeeded in disentangling them again, they believed they had found a deep clue about the fundamental dual existence of the electron and proton. But whatever way one mixes up linearly two Dirac equations, the masses remain just free parameters. Bargmann's article implied that Einstein and Mayer could just as well have written two independent Dirac equations with two different values for

[76] We can also think of the reduction procedure, where a c-transformation is performed on a semivector of the first kind (5.53), as a transformation that exclusively rotates the inert index, i.e. puts any element $g \in SL(2, \mathbb{C})$ in the right-handed representation in (5.51). By definition, however, this is not a Lorentz transformation.
[77] "[D]ie allgemeinsten Semivektor-Gleichungen einfachster Art," (Einstein and Mayer, 1933b, p. 615).
[78] Einstein observed in a letter to Max von Laue (see footnote 53) that the mass ratio had not been fixed by the theory; this is a consequence of the fact that the semivector Dirac equation is just a linear combination of two spinor Dirac systems.

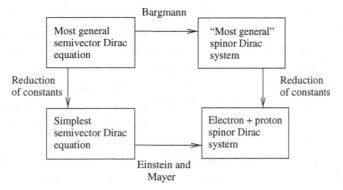

Figure 5.3. It is evident from Bargmann's analysis that the most general semivector Dirac system of Einstein and Mayer is just a linear superposition of two independent Dirac spinor systems. Thus, it does not give insight into the dual occurrence of electrons and protons.

the masses; this would have told as much about the dual nature of electrons and protons as in the end the semivector description had accomplished. There was no reason to expect that the semivector Dirac equation contained any clues about the combined occurrence of electrons and protons.

Regrettably, we have not been able to locate any reaction of Einstein to Pauli's letter, nor to Bargmann's paper. But he did drop the subject around the same time as Bargmann's article appeared, so even if Einstein had not read the article itself, he had by then probably become aware of the superfluous character of the semivector and had lost faith in having found an explanation for the double occurrence of the electron and proton.

5.4 Conclusion: nature and mathematical naturalness

The Dirac equation was at the center of interest of theoretical physicists when Einstein introduced the semivector. What initially started out as the equation's biggest mystery – its negative mass solution, interpreted by Dirac as a "hole" with positive mass and charge in 1930[79] – became its biggest success when the positron was observed a year later. The Dirac equation has remained the central tenet of the now mature theory of quantum electrodynamics, a quantum theory, of course. Einstein however did not engage the Dirac theory as a quantum theory and stood firmly apart from the main lines of development.

In 1932 the neutron was discovered; this particle was not contained in the Dirac equation, neither in spinor nor in semivector form. We believe that Mayer had

---

[79] See "A theory of electrons and protons" Dirac (1930b).

become so strongly convinced of the correctness of the semivector unification of electrons, protons and positrons that he was likely willing to dismiss the neutron altogether – Einstein, however, did not agree with Mayer's cavalier attitude regarding neutrons.[80] Yet positrons and neutrons played hardly any part in Einstein's study of the Dirac equation, nor, we expect, would any other empirical result have easily done so. We have seen that Einstein rather looked to his epistemological convictions and his experiences in the formulation of the general theory of relativity when he was searching for direction and needed to motivate his choices. He further believed that the virtue of these convictions had been corroborated when he had found a unified description of electrons and protons. This result was a fresh triumph of his methodology, which is why it could serve as such an outstanding example in his lecture at the University of Oxford, delivered while the semivector project was at its apex.

That Einstein took his cue from a striving for mathematical naturalness rather than from observation also surfaces in the correspondence that he entertained with his coworker and his colleagues. Paul Ehrenfest raised Einstein's interest in the Dirac equation and spinor theory, not by pointing to its enormous potential with regard to unravelling the behavior of the electron, but instead by arguing that the mathematics of the theory was rather un-intuitive. Einstein's collaborator, Walther Mayer, had indeed been chosen for his qualities as a mathematician, and his background thus suited this project well.

Wolfgang Pauli did not shy away from expressing his dislike of the semivector to Einstein. He had already on earlier occasions confronted Einstein with discrepancies that the latter's theories would have with experience, and with his repeated criticism he likely hoped to induce Einstein to reconsider his attitude to the quantum theory. Besides announcing Valentin Bargmann's critical study, Pauli now pointed to unphysical electron and proton interference terms in the charged semivector current; an unjust critique, but it reaffirms the general pattern of his interventions. Pauli's criticism implicitly confronted Einstein with the successes of the empiricist methods employed by the quantum theorists, as opposed to the failures of his own field theory program.

Einstein had aimed at simplifying and rendering more natural the mathematics of the Dirac equation; yet, his contemporaries found that his efforts made the theory more complex rather than simpler. His project for the Dirac theory in the end failed, and his claim to have enhanced the understanding of the electron and proton could

---

[80] "[...] what I did not understand in your letter is that you think you could deal that easily with the neutrons." ("Das Einzige, was ich in Ihrem Brief nicht verstehe, ist, dass Sie sich so einfach mit den Neutronen abfinden zo können glauben.") Einstein to W. Mayer, 4 June 1933, EA 18-176.

not be upheld. The encounter with semivectors has clearly shown the important role that Einstein's methodological convictions played, as they interacted with his efforts. These convictions reverberated with lessons learnt through the discovery of general relativity. That emotional conquest – "one of the most exciting and demanding times of my life"[81] – had become Einstein's premier reference point.

---

[81] Einstein to A. Sommerfeld, 28 November 1915, Doc. 153, pp. 206–209, on p. 206 in Schulmann *et al.* (1998).

6

Unification in five dimensions

We turn our attention again to Einstein's attempts to unify gravity and electro-magnetism; in particular, we intend to address various aspects of his study of the Kaluza–Klein theory. Einstein studied this theory extensively on a number of occasions. His deepest involvement was in the late 1930s and early 1940s, producing two substantial papers. By this time Einstein had become firmly settled at the Institute for Advanced Study, where his Kaluza–Klein collaborators, Peter Bergmann and Valentin Bargmann, were employed as his assistants – again a short introduction to both of them is contained in the boxed texts of this chapter.

6.1 Particle solutions in field theory

Einstein was quite productive in the years between the semivector and these later Kaluza–Klein publications. In 1935, the "Einstein–Podolsky–Rosen" paper on the completeness of quantum mechanics appeared, and in 1938 he co-authored another important article on "the problem of motion" – the relation between the geodesic equation for test particles and general relativity's field equations; he collaborated with Leopold Infeld and Banesh Hoffmann on this subject. Einstein published a study on gravitational lensing in 1936 and the same year he and Nathan Rosen submitted an article to the *Physical Review* that called into question the existence of gravitational waves. The journal's referee objected; this was quite likely a new experience for Einstein – peer review not being common in the German speaking world – and he decided to publish the paper in the less prominent *Journal of the Franklin Institute* instead, though with altered conclusions. As it turned out, Einstein would never publish in the *Physical Review* again.[1]

---

[1] See Einstein and Rosen (1936); for more on this episode, see Kennefick (2007), pp. 79–104. Einstein's most important paper on the problem of motion is Einstein *et al.* (1938). Finally, let us point out that recent scholarship has established that Einstein had in fact already thought of gravitational lensing while he was still involved with the Entwurf theory (Renn *et al.*, 1997); his first publication on the subject was Einstein (1936b).

These publications, despite their lasting impact, were however not much more than sidelines to Einstein's larger research program, with its ultimate goal of re-deriving from classical field theories the quantum nature of matter. Before taking up this issue again in Kaluza–Klein theory, Einstein had studied what general relativity might have to offer in this respect. Together with Rosen he formulated what is now generally referred to as the Einstein–Rosen bridge, originally proposed as an attempt to unify field and particle concepts.[2]

Let us briefly describe the Einstein–Rosen bridge. Einstein and Rosen had glued together two static rotation symmetric solutions at their Schwarzschild radii and had dismissed the part of spacetime that lies beyond that radius. A coordinate trans-formation was performed which put the coordinate singularity of the Schwarzschild radius in the origin of a new coordinate system, and regularized it. Finally, they proposed to alter the field equations in such a way that the singularity did not show up in the new field equations either. The two connected, curved spacetime "sheets" were proposed as a "bridge"-like model for a massive particle.[3]

The Einstein–Rosen bridge exemplifies how Einstein often attempted to inte-grate atomicity and elementary particles into a classical field theory: the classical field equations should give a non-singular solution, whose dimensions and interac-tions should be governed by quantal relations. These were also to follow from the classical field equations. In the case of the gravitational bridge solution, the idea was that it could perhaps serve as a model for the neutron since it did not carry an electromagnetic charge. Einstein was quite positive about the possibilities that the bridge might offer: in his 1936 essay "Physics and reality," he pointed out that the bridge was essentially a "discrete element" and that if he could succeed in finding a multi-bridge solution of the field equations, this could perhaps explain the equality of the elementary masses found in nature.[4]

We have seen that already at the inception of the unified field theory program Pauli identified the central role played by singularity-free particle solutions. Indeed, Einstein tried to find such solutions in most of the classical field theories that he stud-ied, but all his attempts eventually failed. Also his assessment of the Einstein–Rosen bridge as a non-singular solution was unfortunate, for, like most of his contempo-raries, Einstein had overlooked the perspective of an observer that falls towards the Schwarzschild radius.[5] Such an in-falling observer can pass the Schwarzschild

---

[2] See Einstein and Rosen (1935).

[3] In the new coordinates $g^2 R_{\mu\nu}$ was not singular. Consequently, Einstein and Rosen proposed to multiply the gravitational field equations with g^2. See Einstein and Rosen (1935), "sheet" and "bridge" on its p. 75. For historical commentary, see Earman and Eisenstaedt (1999), pp. 213–220.

[4] See Einstein (1936a), on p. 354 in Einstein (1994).

[5] Einstein would also do so in a paper that studied geodesic motion in Schwarzschild spacetime, see Einstein (1939). For more on the confusion regarding the Schwarzschild horizon in early studies of general relativity, see Eisenstaedt (1982, 1987, 1989a, 1993).

radius unscathed – the singularity there being only a coordinate artefact – into the region that Einstein and Rosen had believed they could do without. The observer would end up crashing into an intrinsic spacetime singularity that is located at the origin of the Schwarzschild coordinates; this implies that the Einstein–Rosen bridge cannot serve as a non-singular particle solution. Some in Einstein's vicinity were in fact aware of the spurious character of the singularity at the Schwarzschild radius, in particular the Princeton cosmologist Hartland P. Robertson. It has, however, not yet been possible to establish to what extent Einstein was informed about these ideas;[6] in any case he soon decided to abandon the bridge as a non-singular particle model.

The construction of a singularity-free particle solution was again a pivotal element in the elaboration of the Kaluza–Klein theory that Einstein began in 1937. Before we follow his efforts in more detail, we start by giving a historical introduction of the theory by presenting the original work of Theodor Kaluza and Oskar Klein from the 1920s. The papers of Kaluza and Klein inspired two publications by Einstein, and we will also briefly describe these. We omit the short episode in which he studied a projective formulation of the Kaluza–Klein theory with Walther Mayer in the fall of 1931 and winter of 1932, largely because, already at an early stage of their project, Einstein and Mayer expressed themselves pessimistic regarding the possibility of achieving the desired "understanding of the nature of corpuscules or [...] the results established in quantum mechanics."[7] But five years later, Kaluza–Klein again looked like a promising candidate.

6.2 Kaluza, Klein and Einstein

In 1919, Theodor Kaluza, a *Privatdozent* in mathematics in Königsberg, communicated to Einstein the idea of studying a five-dimensional Riemannian geometry as a unified field theory. Einstein initially quite liked Kaluza's suggestion and encouraged him to elaborate it further. Kaluza's publication would be delayed by two years, however, in part because he and Einstein observed a particular shortcoming, which they expected to be resolved by further study (we will see shortly what the shortcoming entailed). Eventually Kaluza's paper "Zum Unitätsproblem der Physik" ("On the unity problem in physics") was presented by Einstein to the Prussian Academy in 1921.[8]

---

[6] Nevertheless, it is known that Einstein did regularly discuss relativity with Robertson. Robertson was well aware of the causal structure of the Schwarzschild solution by 1936 or 1937, as follows from the posthumous publication of his lecture notes of those years (Roberston and Noonan, 1968). For more on this issue, see Eisenstaedt (1987), pp. 328–338.

[7] The two articles containing Einstein and Mayer's work on the projective Kaluza–Klein theory are Einstein and Mayer (1931, 1932a); on their pessimism, see Vizgin (1994), pp. 258–261, quotation of Einstein and Mayer on p. 260.

[8] Parts of this chapter appeared earlier in van Dongen (2002). For Einstein's correspondence with Kaluza, see the letters reprinted in Buchwald *et al.* (2004, 2009); Kaluza's paper is Kaluza (1921); more on his contribution can be read for example in Goenner and Wünsch (2003), Wünsch (2003), and Goenner (2004), section 4.2.

Kaluza's article started by noticing a formal similarity between the definitions of the electromagnetic field, in terms of its potentials, and the Christoffel symbols from gravitation theory in terms of the metric; perhaps the former could somehow be truncated manifestations of the latter? Kaluza argued that in a five-dimensional theory the electromagnetic field indeed could be taken to be exactly that, if one imposed a "cylinder condition," that is, if one required that the components of the five-by-five metric g_{IJ} do not depend on the space-like fifth dimension. He then interpreted the five-by-five metric as follows (here capital Latin indices run over 0, 1, 2, 3, 5; Greek indices run over the usual dimensions 0, 1, 2, 3):

$$g_{IJ}^{(5)} = \begin{pmatrix} g_{\mu\nu}^{(4)} & 2\alpha A_\nu \\ 2\alpha A_\mu & 2V \end{pmatrix} \quad \text{with} \quad \partial_5 g_{IJ} = 0, \tag{6.1}$$

where $g_{\mu\nu}$ represents the metric of the four-dimensional spacetime, A_μ is the gauge field from electrodynamics and α is a constant, related to the Newton constant κ via $\alpha = \sqrt{\kappa/2}$. The theory contained an additional new scalar field, $V(x)$.

Kaluza evaluated the Ricci tensor for linearized fields, finding:

$$\begin{aligned} R_{\mu\nu} &= \partial_\lambda \Gamma^\lambda_{\mu\nu}, \\ R_{5\nu} &= -\alpha \nabla^\mu F_{\mu\nu}, \\ R_{55} &= -\Delta V. \end{aligned} \tag{6.2}$$

He continued by studying the example of charged matter in the five-dimensional space. This revealed that the momentum in the fifth direction needed to be interpreted as the electric charge, and that generally $j^\nu = 2\alpha T^{\nu 5}$; charge was further a conserved quantity.

From the five-dimensional line element Kaluza recovered for the equation of motion of a charged particle:

$$m\left(\frac{d^2 x^\mu}{d\tau^2} + \Gamma^\mu_{\beta\gamma} \frac{dx^\beta}{d\tau} \frac{dx^\gamma}{d\tau}\right) = 2\alpha e \, F^\mu{}_\nu \frac{dx^\nu}{d\tau} - \frac{e^2}{m} \partial^\mu V. \tag{6.3}$$

This complies with the Lorentz force law under the assumption of small specific charge e/m. For the electron, however, this assumption breaks down and contrary to experience, the interaction with the scalar field becomes the leading term, as was pointed out to Kaluza by Einstein; this was the problem that led to the initial delay in the publication of Kaluza's work.[9] By 1921, however, both Einstein and Kaluza had become less impressed by it, and agreed that a publication was warranted.

The geometric theory that Kaluza initially formulated was – until he introduced matter source terms in his example – a five-dimensional vacuum theory. The field equations would be given by equating the Ricci tensor (6.2) to zero. Yet, once a

---

[9] See Einstein to Theodor Kaluza, 14 May 1919, Doc. 40 in Buchwald *et al.* (2004); Kaluza (1921), pp. 971–972.

particle or other matter terms are explicitly introduced one has left the vacuum theory; when Kaluza in his example wrote $R_{MN} = \kappa T_{MN}$, the source term had no origin in his five-dimensional unification geometry. Einstein addressed this point in a 1923 paper that he wrote with his assistant of the time, Jakob Grommer:

Herr Kaluza introduces besides the quantities g_{IJ} another tensor representing the material current. But it is clear that the introduction of such a tensor is only intended to give a provisional, sheer phenomenological description of matter, as we presently have in mind as ultimate goal a pure field theory, in which the field variables produce the field of "empty space" as well as the electrical elementary particles that constitute "matter."[10]

Einstein wanted the sources to come from the geometry, and in the paper with Grommer he intended to see whether he could find non-singular particle solutions proper to the vacuum Kaluza field equations themselves. Then one would no longer need to introduce separate matter terms and Kaluza's theory would have achieved not just a unification of electromagnetism and gravity, but a unification of field and particle as well.

In their opening paragraphs, Einstein and Grommer explained their enthusiasm for Kaluza's theory: the five-dimensional unification was of a "staggering formal simplicity," principally because the gravitational and electromagnetic fields were no longer two disparate terms, "outwardly welded together by a plus sign."[11] However, the theory also had its shortcomings. Firstly, the five-dimensional line-element lacked a straightforward operational definition, in contrast to the four-dimensional line-element which with rods and clocks could be measured directly. Secondly, the cylinder condition singled out one dimension, introducing an undesirable asymmetry to the theory. The conclusion of the paper, however, gave the theory's greatest failing: Einstein and Grommer found that the vacuum field equations ($R_{IJ} = 0$) could not produce non-singular rotation symmetric particle solutions.

This meant that Kaluza's theory could not give the desired description of the charged matter sources. It seems highly likely that for this reason Einstein dropped the theory. He wrote to Hermann Weyl:

[Kaluza] seems to me to have come closest to reality, even though he too fails to provide the singularity free electron. To allow singularities does not seem to me the right way. I think that in order to truly make progress one must once more ferret out a general principle from nature.[12]

---

[10] "Herr Kaluza führt allerdings ausser den Grössen $g_{\mu\nu}$ noch einen Tensor der materiellen Strömung ein. Aber es ist klar, dass die Einführung eines solchen Tensors nur dazu dient, eine vorläufige, bloss phänomenologische Beschreibung der Materie zu geben, während uns heute als letztes Ziel eine reine Feldtheorie vorschwebt, derart, dass die Feldvariabeln sowohl das Feld des „leeren Raumes" als auch die elektrischen Elementarteilchen darstellen, welche die „Materie" ausmachen" (Einstein and Grommer, 1923, p. 2).

[11] "[V]erblüffender formaler Einfachheit" – "[...] die durch das Plus-Zeichen äusserlich zusammengeschweisst sind." As in Einstein and Grommer (1923), pp. 1–2.

[12] "[Kaluza] riecht mir noch am ehesten nach Realität, aber das singularitätsfreie Elektron liefert er auch nicht. Singularitäten zuzulassen scheint mir nicht der richtige Weg. Ich glaube, man müsste – um wirklich vorwärts zu kommen, wieder ein allgemeines, der Natur abgelauschtes Prinzip finden." Einstein to Hermann Weyl, 6 June 1922, EA 24-71.

Kaluza's intuitive leap to five dimensions had been a very appealing idea, but alas, it did not produce the desired description of charged matter. Now that five-dimensional relativity had apparently failed, some new organizing principle needed to be found. Einstein next moved to the theory that had been formulated in 1921 by Arthur S. Eddington, a theory in which the affine connection rather than the metric was the fundamental quantity (this was not the same theory as the Eddington theory which Einstein dismissed during the semivector episode).[13] He briefly returned to the five-dimensional theory in 1927, after the work of Oskar Klein had appeared.

The Swede Oskar Klein had at first not been aware of Kaluza's publication, and thought the five-dimensional theory through afresh.[14] Unlike Kaluza, Klein involved quantum relations in his articles: this taught him that the fifth direction can be thought of as a compact dimension. He published one short note on the five-dimensional theory in *Nature* and a more elaborate paper in the *Zeitschrift für Physik*.[15] Both articles came out in 1926, after the Schrödinger equation had been published.

The German article was divided into a restatement of Kaluza's classical theory and a study of the possibility of relating the quantum mechanical wave equation with five-dimensional (null-)geodesics. In the first part Klein put forward a more fruitful interpretation of the five-dimensional metric, constrained by the cylinder condition:

$$g_{IJ}^{(5)} = \begin{pmatrix} g_{\mu\nu}^{(4)} + VA_\mu A_\nu & VA_\nu \\ VA_\mu & V \end{pmatrix}. \tag{6.4}$$

Klein put the scalar V to a constant; he argued that this was permitted as g_{55} would remain invariant under general coordinate transformations of the regular four spacetime dimensions and translations in the fifth dimension, the transformations that complied with the cylinder condition. He pointed out that the theory contained electromagnetic gauge transformations as translations of the fifth dimension:

$$x^5 \to x^5 + \Lambda(x^\mu), \qquad A_\mu \to A_\mu + \partial_\mu \Lambda. \tag{6.5}$$

Klein introduced the action of his theory as the Ricci scalar for the five-dimensional metric. He subsequently varied this action and retrieved the full – not merely linearized – field equations of general relativity with the energy-momentum tensor of

---

[13] On Eddington's 1921 theory and Einstein's interest in it, see Vizgin (1994), pp. 137–149, Goenner (2004), section 4.3.
[14] For more on Klein, see for example Pais (2000), pp. 122–147; see also Halpern (2007). For commentary on Klein's article, see Goenner and Wünsch (2003).
[15] See Klein (1926a,b).

the electromagnetic fields, together with the source free Maxwell equations:[16]

$$G_{\mu\nu} = \kappa T_{\mu\nu}, \qquad \frac{\partial \sqrt{-g}F^{\mu\nu}}{\partial x^{\nu}} = 0. \tag{6.6}$$

Klein further claimed that the theory now also reproduced the equation of motion for charged particles "in the usual way," i.e. by variation of the five-dimensional line element.[17] It of course did not contain the disturbing scalar field contribution from Kaluza's analysis.

In the second part of his *Zeitschrift für Physik* article, Klein introduced charged matter terms as massless wave fields in five dimensions. He proposed a generalized and covariant wave equation:

$$a^{IK}\left(\frac{\partial^2 U}{\partial x^I \partial x^K} - \Gamma^R_{IK}\frac{\partial U}{\partial x^R} \right) = 0, \tag{6.7}$$

and subsequently chose a metric for a_{IK} that would leave only mass and electrostatic terms to remain in the above expression. Klein then made the Ansatz

$$U = e^{-2\pi i \left(\frac{x^5}{h} - \nu t \right)} \psi(x, y, z) \tag{6.8}$$

with which he could obtain the Schrödinger equation.

The above Ansatz reveals that Klein originally attributed the fifth dimension the same units as Planck's h. In the second short note on his proposal, however, it had acquired the unit of length and Klein stated that if charges can only have discrete values, as they do in nature,

$$p_5 = n\frac{h}{\lambda_5} = \frac{ne}{c\sqrt{2\kappa}}, \tag{6.9}$$

the fifth dimension needs to be compact, i.e. it would be periodic with period λ_5. With (6.8) one sees that translations $x^5 \to x^5 + \Lambda(x)$ correspond to gauge transformations of the wave-function.

In Klein's theory, the gauge parameters implicitly took on values on the circle, implying, in modern terminology, that the gauge group of electrodynamics should be the compact $U(1)$ group. The topology of space has changed as the fifth direction is no longer represented by an infinite straight line, but must be thought of as a circle. The above also means that the scale invariance of the fifth dimension has

---

[16] The inconsistency that follows from putting the scalar V to a constant *after* varying, as noted by for instance Yves Thiry (1948), thus does not surface. The field equation for the scalar would then impose $F^2 = 0$. For more on the role of the scalar field, see also O'Raifeartaigh and Straumann (2000), p. 10.

[17] As in Klein (1926b), cited from p. 63 in O'Raifeartaigh (1997).

been restricted: an infinite straight line can be scaled arbitrarily, but a compact fifth dimension implies that the scale of that dimension is fixed, here by equation (6.9).

Putting in the values of the parameters revealed that the scale of the fifth direction is very small, close to the Planck size:

$$\lambda_5 = \frac{hc\sqrt{2\kappa}}{e} = 0.8 \times 10^{-30}\,\text{cm}. \tag{6.10}$$

The invocation of a fifth dimension could now be considered a less arbitrary assumption, as Klein observed:

The small value of this length together with the periodicity in the fifth dimension may perhaps be taken as a support of the theory of Kaluza in the sense that they may explain the non-appearance of the fifth dimension in ordinary experiments as the result of averaging over the fifth dimension.[18]

The paper that Einstein published shortly after Klein's essentially only reproduced the latter's enhanced representation of the classical Kaluza theory (which Einstein acknowledged in an appendix), and he again gave it a positive assessment.[19] Einstein's commentary on the five-dimensional unification once more exhibited a close link between unification and understanding: "one may say that in the framework of the general theory of relativity Kaluza's idea provides a rational foundation for the electromagnetic equations of Maxwell and unifies these with the gravitational equations into a formal whole."[20]

The most revealing aspect of Einstein's article lies in the fact that it did not make any mention of a possible compactification of the fifth dimension, nor of the quantization of charge: Einstein did not comment on the part of Klein's papers that connected the five-dimensional theory to quantum relations, despite the fact that he was well aware of Klein's treatment of this issue.[21] We feel that the reason for this neglect lies in Einstein's desire for a field theory that could *derive* the discreteness of charge. Klein on the other hand had basically accepted this discreteness as a given fact; Einstein may have felt that Klein's scheme had not given an actual *explanation* of the charge discreteness. Furthermore, since Klein had had to add the charged sources as additional wave-fields – that had no obvious geometrical origin – these had not been unified with the other fields, and were therefore essentially still not understood.

[18] As in Klein (1926a), p. 516.
[19] In his appendix, Einstein (1927), p. 30, further mentioned work by Vladimir Fock (1926) that contained results similar to Klein's.
[20] "[Man] kann sagen, dass Kaluzas Gedanke im Rahmen der allgemeinen Relativitätstheorie eine rationelle Begründung der Maxwellschen elektromagnetischen Gleichungen liefert und diese mit den Gravitationsgleichungen zu einem formalen Ganzen vereinigt," (Einstein, 1927, p. 30).
[21] See on this awareness Halpern (2007), pp. 395–399.

Before turning to the Einstein–Bergmann–Bargmann elaboration of the five-dimensional theory, it should be helpful to look at some current ideas on how the charged sources can be incorporated in Kaluza–Klein theory. In a compactified five-dimensional theory one can expand the components of the metric in a Fourier series:

$$g_{IK} = \sum_n g_{nIK}(x^\mu) e^{in\frac{x^5}{\lambda_5}}. \tag{6.11}$$

One generally finds this decomposition in more recent literature.[22] $g_{0IK}(x^\mu)$ is just the metric given in (6.4) and the higher Fourier terms, that can carry a momentum in the fifth direction, can be identified with charged sources.

In this setup, assuming the discreteness of charge implies the quantization of the field theory, i.e. it implies that one needs to turn continuous parameters into counting operators. We would like to illustrate this briefly by turning to a compactified two-dimensional space upon which a massless field ϕ lives, representing higher order contributions to the metric:

$$\phi(x^0, x^5) = \sum_n \frac{a_n}{\sqrt{\omega_n}} e^{i(n\frac{x^5}{\lambda_5} - \omega_n x^0)}. \tag{6.12}$$

Charge is the component of momentum in the x^5-direction:

$$P^5 = \int dx^5 \; T^{05} = \int dx^5 (\partial^0 \phi)^* \partial^5 \phi. \tag{6.13}$$

From this follows:

$$P^5 = \frac{1}{\lambda_5} \times \sum_n n a_n^* a_n. \tag{6.14}$$

The coefficients a_n^* and a_n are ordinary complex numbers that can take a continuous range of values, so in a classical theory the compactification does not give a discrete spectrum for the charge. For this, the theory needs to be quantized. In other words, imposing the de Broglie relation with integer charges – as Klein had done – should imply that the continuous coefficients a_n^* and a_n are turned into creation and annihilation operators.

Upon quantization one retrieves from the five-by-five metric (6.11) a "tower" of charged and massive spin-2 particles. With Klein's value for the scale of the fifth dimension, their masses would be very high as these would be inversely proportional to that scale. As effective theory for low energy physics, one can retain just the $n = 0$

----

[22] See for instance Duff *et al.* (1986), p. 17.

mode in the Fourier expansion. This would give for the action of the metric:

$$R_0^{(4)} + \frac{1}{4}V_0 F_{0\mu\nu}F_0^{\mu\nu} - \frac{2}{V_0^{1/2}}\Box V_0^{1/2} \tag{6.15}$$

which is the action of the theory with the cylinder condition in place. In future, we will refer to this theory as *Kaluza's theory*, even though Kaluza only studied a linearized version of it. The theory with higher Fourier components – in other words, with periodic x^5-dependence – we will call *Kaluza–Klein theory*. Note that the Kaluza theory has "forgotten" about the periodicity in x^5 of the full Kaluza–Klein theory.

6.3 Einstein–Bergmann–Bargmann theory

We now make a leap from the time that Kaluza and Klein's articles appeared and turn to Einstein's studies of these theories from the late 1930s and early 1940s. The first paper of this period, written together with Peter Bergmann (Figure 6.1), started with a re-evaluation of Kaluza's theory in which the scalar field V had been set to a constant. Einstein and Bergmann argued that the theory had a number of shortcomings; the cylinder condition, for instance, was deemed "artificial." They further reported that:

Many fruitless efforts to find a field representation of matter free from singularities based on this theory have convinced us [...] that such a solution does not exist. [footnote:] We tried

Figure 6.1. Peter Bergmann in 1955 (Syracuse University, courtesy AIP Emilio Segrè Visual Archives).

Einstein and the education of Peter Gabriel Bergmann

Peter Bergmann (1915–2002; see Figure 6.1) was born in a liberal Jewish-German academic milieu. His father Max was a noted organic chemist and acquaintance of Einstein, his mother Emmy was a pediatrician and formed her own Montessori *Kinderhaus* in 1922, which was attended by Peter. His parents soon came to believe that their son was an extraordinarily gifted child – indeed, at the age of sixteen he started his university studies and at the age of eighteen he was ready to do his doctoral work.[23]

In the summer of 1933, Emmy Bergmann wrote to Einstein concerning Peter's graduate education. She was hoping that Einstein might consider taking her son on as a doctoral student. Bergmann had intended to do his graduate studies in Berlin, but his mother feared that this would be impossible because of the political situation.[24] Einstein replied that he was impressed by the apparent intellectual gift of young Peter. His advice was that Bergmann should finish his education at the Zurich ETH, where the "very talented and still young professor Pauli lectures."[25] Einstein explained that he himself was about to leave for Princeton, but he would consider inviting Bergmann to the Institute for Advanced Study as an assistant once he had finished his doctoral thesis. Three years later, in June 1936, Peter Bergmann defended his thesis – not at the ETH but in Prague, where he had studied under Philipp Franck. He had undertaken a mathematical study of the Schrödinger equation on a sphere. Bergmann sent a copy of his thesis to Einstein, and asked if he indeed could continue his studies as his assistant:

It is clear to me that I still have much to learn, in particular with respect to the general theory of relativity on the one hand, and the quantum theory on the other hand. If possible, I would very much like to continue directing my work at finding the link between these two fields.[26]

to find a rigorous solution of the gravitational equations, free from singularities, by taking into account the electro-magnetic field. We thought that a solution of a rotation symmetric character could, perhaps, represent an elementary particle. Our investigation was based on the theory of "bridges" [...]. We convinced ourselves, however, that no solution of this character exists.[27]

The version of the Kaluza theory considered here only contained the usual gravitational and electromagnetic fields. Since this theory apparently could not produce particle solutions, Einstein and Bergmann opted for a theory that included more fields: they moved to a fuller Kaluza–Klein type theory.

---

[23] For a more extensive account of Peter Bergmann's life, see Halpern (2005); see also his obituary in the *New York Times*: Dennis Overbye, "Peter G. Bergmann, 87; Worked with Einstein," 23 October 2002.

[24] Emmy Bergmann to Einstein, July 1933, EA 6-220.

[25] "[...] sehr begabte und noch junge Professor Pauli lehrt." Einstein to Emmy Bergmann, 5 August 1933, EA 6-221.

[26] "Ich bin mir darüber klar, dass ich besonders auf dem Gebiet der Relativitätstheorie einerseits, der Quantenmechanik anderseits, noch sehr viel zu lernen habe. Wenn es mir möglich wäre, würde ich sehr gern in der Richtung weiterarbeiten, die Verbindung zwischen diesen beiden Gebieten zu suchen." Peter Bergmann to Einstein, 14 March 1936, EA 6-222.

[27] As in Einstein and Bergmann (1938), p. 688, "artificial" on p. 696.

When Einstein replied to Bergmann, he again expressed his amazement at the fact that Bergmann had achieved such intellectual maturity at his young age – he was still only 21 years old. Yet Einstein did not eschew telling him that he did not think the thesis addressed truly interesting issues. Nevertheless, Einstein would take Bergmann on as an assistant, and believed that, given the latter's interests, they should appreciate working together. Soon Einstein and Bergmann were studying the Kaluza–Klein theory. Their collaboration lasted for some four years and resulted in two papers.[28]

Among Bergmann's lasting contributions ranks his 1942 *Introduction to the Theory of Relativity*, published with a foreword by Einstein.[29] It is a text book at the graduate level, which steadily advanced to the topics that were studied at the time by Einstein and his assistants. Its last two chapters were devoted to their elaborations of the Kaluza–Klein theory.

In one of his numerous letters of reference on Bergmann's behalf (finding a job in the United States in the early forties was hard due to the influx of European refugees), Einstein wrote that Bergmann was "by no means limited to my own special interests."[30] Evidently, it was no longer a positive recommendation to be too strongly involved with Einstein's field theory program. Bergmann indeed found his intellectual independence and did not restrict himself to Einstein's unified field theory approach: he was among the first to attempt to quantize the gravitational interaction and initiated the "canonical" formulation of general relativity.[31]

The fact that spacetime appears to us as four-dimensional was again proposed, as in Klein's work, to be due to a compactification of the fifth dimension. To elucidate the argument, Einstein and Bergmann gave the example of a compactified two-dimensional space (see also Figure 6.2):

If the width of the strip, that is, the circumference of the cylinder (denoted by b), is small, and if a continuous and slowly varying function $\phi(x^5, x^1)$ is given, that is if $b\partial\phi/\partial x^a$ is small compared with ϕ, then the values of ϕ belonging to the points on the segment PP' [i.e. the line connecting the periodically identified points P and P'] differ from each other very slightly and ϕ can be regarded, approximately, as a function of x^1 only.[32]

In this example, the dimension x^1 was the analog of the familiar four dimensions of spacetime. The function ϕ could be expanded as in (6.12) and, analogously, Einstein and Bergmann could have written the five-dimensional metric as in (6.11), but they did not explicitly do so, in spite of the assumed periodicity in the fifth direction. In any case, they took the scale of the extra and compact direction to be small, and in the vocabulary of the Fourier decomposed metric, one could say that Einstein and Bergmann further tacitly made the assumption that the $n \neq 0$ Fourier

---

[28] See Einstein to Peter Bergmann, 7 June 1936, EA 6-226; their papers are Einstein and Bergmann (1938); Einstein *et al.* (1941).

[29] See Bergmann (1942). In 1975 a Dover edition appeared which contained two new appendices.

[30] Einstein to Vern O. Knudsen, dean of the UCLA physics department, 1 June 1940, EA 6-287.

[31] See his recollections in Bergmann (1992).

[32] As in Einstein and Bergmann (1938), p. 689.

Bergmann started his teaching career at Black Mountain College in North Carolina. In 1949 he settled at Syracuse University, where he again wished to study how gravity and the quantum could be reconciled. In 1954, Bergmann asked the National Science Foundation for financial support of his research. The NSF forwarded his request for review to Einstein, who replied:

The application of Dr. Bergmann concerns a problem of central significance for modern physics. All physicists are convinced of the high truth value of the probabilistic quantum theory and of the general relativity theory. These two theories, however, are based on independent conceptual foundations, and their combination to a unified logical system has so far resisted all attempts in this direction. [...] If the decision were mine I should grant the funds asked for by Dr. Bergmann, in view of the central importance of the problem and the qualifications of the candidate. Even though the probability of attaining the great goal seems rather small at this point, the financial risk incurred is on the other hand so modest that I should have no qualms to grant the application.[33]

Bergmann's group in Syracuse became one of the leading centres for gravitational research in the United States.

Bergmann, unlike Walther Mayer, may not have had much difficulty in adjusting to life in America; he was involved in public causes – for instance the protest against the intake by the United States of scientists that had worked for National-Socialist Germany[34] – and in his letters to Einstein the number of English words that crept into his German steadily increased. Peter Bergmann died in Seattle on October 20, 2002.

coefficients of the metric were small relative to the $n = 0$ coefficient – this would explain how the world appears to us as if there are four dimensions instead of five.

The theory encompassed the original symmetries of gravity and electromagnetism and reduced to Kaluza's theory if the x^5-dependence was dropped.[35] Einstein and Bergmann's richer geometry however retained x^5-dependent terms and thus also incorporated charged matter sources. They proposed the following set of field equations:

$$\alpha_1 \left(R_{\mu\nu} - \frac{1}{2} g_{\mu\nu} R \right) + \alpha_2 \left(2 F_{\mu\alpha} F_\nu^\alpha - \frac{1}{2} g_{\mu\nu} F^{\alpha\beta} F_{\alpha\beta} \right)$$

$$+ \alpha_3 \left(-2 g_{\mu\nu,55} + 2 g^{\alpha\beta} g_{\mu\alpha,5} g_{\nu\beta,5} - g^{\alpha\beta} g_{\alpha\beta,5} g_{\mu\nu,5} - \frac{1}{2} g_{,5}^{\alpha\beta} g_{\alpha\beta,5} g_{\mu\nu} \right)$$

$$+ \alpha_4 g_{\mu\nu} \left(\frac{1}{2} (g^{\alpha\beta} g_{\alpha\beta,5})^2 + 2 g^{\alpha\beta} g_{\alpha\beta,55} + 2 g_{,5}^{\alpha\beta} g_{\alpha\beta,5} \right) = 0, \qquad (6.16)$$

[33] Einstein to the National Science Foundation, 18 April 1954, EA 6-313.
[34] See Bergmann's letter to James W. Wise of the Council Against Intolerance in America, 16 December 1946, EA 6-281.
[35] See in this regard also Einstein *et al.* (1941), p. 217.

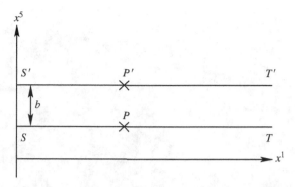

Figure 6.2. After Einstein and Bergmann, who gave the above two-dimensional illustration of the compactified space they studied: "We imagine the strip curved into a tube so that *ST* coincides with *S'T'*." See Einstein and Bergmann (1938), pp. 688, 689.

$$\int dx^5 \sqrt{-g}\,\left\{ \alpha_1(g^{\alpha\beta}\Gamma^{\mu}_{\alpha\beta,5} - g^{\alpha\mu}\Gamma^{\beta}_{\alpha\beta,5}) - 4\alpha_2 F^{\mu\alpha}_{;\alpha} \right\} = 0. \qquad (6.17)$$

The first equation is the Einstein equation with electromagnetic fields and charged matter terms. The last expression (6.17), i.e. Maxwell's equations, shows that the x^5-dependent metric fields indeed played the role of charged sources: the electric current is given by the connection terms.

The α_i were constants, and their value did not appear to be fixed by the theory in the first paper. From his correspondence we learn that Einstein was quite uncomfortable about this element of arbitrariness.[36] His discomfort was quite likely epistemologically motivated: a little later, in his autobiography, he would state a "theorem" that claimed that "there are no *arbitrary* constants of [a dimensionless] kind." The theorem was ultimately only based upon nothing more than Einstein's "faith in the simplicity, i.e. intelligibility, of nature"; he believed that "nature is so constituted that it is possible logically to lay down such strongly determined laws that within these laws only rationally completely determined constants occur."[37]

In the second publication on the theory, co-authored not just by Bergmann but now also by Valentin Bargmann (see Figure 6.3), the value of the constants was fixed – now five-dimensional general covariance was imposed which determined their value. This introduced a new problem however: the electromagnetic and

[36] Einstein to V. Bargmann, 9 July 1939, EA 6-207; see also p. 1 in the unpublished manuscript "Ein Gesichtspunkt für eine spezielle Wahl der in der verallgemeinerten Kaluza-Theorie auftretenden Konstanten," EA 1-136, dated at 1941 by the Einstein Archive.
[37] As in Einstein (1949a), p. 63.

Figure 6.3. Princeton, 1940: Einstein at the blackboard with Bergmann and Bargmann (Lucien Aigner/Corbis).

gravitational interaction were in this case of the same order of magnitude, a strong argument against the empirical validity of this formulation of the theory.[38]

The Einstein–Bergmann–Bargmann theory looked like it followed Klein's logic; Einstein and his assistants now retained the fifth dimension, unlike in Einstein's earlier follow-up paper from 1927, and indeed compactified it. Yet there still was an important difference: the theory was studied exclusively as a classical theory. Einstein did not introduce the value of $\hbar$ as Klein had done when he used the de Broglie relation and the discreteness of the electric charge to determine the scale of the compactified fifth dimension. Without the de Broglie relation and $\hbar$, Einstein could not calculate the value of his scale parameter b (i.e. λ_5), nor could he a priori account for the quantization of charge. He took the scale of the fifth direction, his parameter b, to be "small," but he did not put a number to that smallness.

---

[38] See Einstein *et al.* (1941), p. 225. Note further that five-dimensional general covariance in fact would imply that the compactification would be given up.

Again we think that the reason why Einstein made this omission was that he intended the theory to *explain* the quantized nature of matter: he could not just assume the empirical discreteness of charge, this he wanted to *deduce*. The two publications do not explicitly exhibit that this was one of Einstein's actual goals for the theory, but we have concluded as much from his unpublished writings and his correspondence with Bergmann and Bargmann, to which we shall turn soon.

Einstein's neglect of the de Broglie relation in his Kaluza–Klein publications reflects his conviction that quantum theory was to follow from a classical unified field theory. In the construction of this deeper unified theory, the quantum relations should not be included at the outset:

[Quantum mechanics] will be a touchstone for any future theoretical basis, in that it must be deducible as a limiting case from that basis, just as electrostatics is deducible from the Maxwell equations of the electromagnetic field or as thermodynamics is deducible from classical mechanics. However, I do not believe that quantum mechanics can serve as a *starting point* in the search for this basis, just as, vice versa, one could not find from thermodynamics [...] the foundations of mechanics.[39]

Quantum relations were to follow – "top-down" – from a unified field equation, just as, if we may add another familiar example to Einstein's list, Newtonian gravity had ultimately been deducible from generally relativistic gravity laws.

In Einstein's efforts, the de Broglie relation from quantum theory would thus not "serve as a starting point." For him, the quantum relations were not much more than empirical regularities, mere phenomenological laws – one could say that they were confined to the "E," or, at best, "S"-level in the Solovine schema (we will return to this in Chapter 7). Klein, however, by invoking the de Broglie relation at the outset of his reasoning, had allowed a "bottom-up" argument to creep into his logic; his line of reasoning exhibited the dual character – combining familiar physical relations with five-dimensional mathematical theory – that reminds one of the dual Einstein from before November 1915. Yet, in the 1930s Einstein held that the unification of electromagnetism and gravity should begin with a new and mathematically convincing set of axioms, from which quantum relations were to follow, like the S_7 and a new match with E follow from the unifying A' in Figure 3.1. The Einstein–Bergmann–Bargmann version of Kaluza–Klein theory was the latest structure under scrutiny to that end.[40]

---

[39] See Einstein (1936a), on p. 352 in Einstein (1994).

[40] In an unpublished manuscript on the theory, Einstein indeed wrote: "The theory presented in the preceding gives a formally fully satisfying unified conception of the structure of physical space. Further investigations will have to show whether it contains a theory of elementary particles as well as of the quantum phenomena (free from statistical elements)." ("Die im Vorstehenden entwickelte Theorie gibt eine formal völlig befriedigende einheitliche Auffassung von der Struktur des physikalischen Raumes. Wietere Untersuchungen müssen zeigen, ob sie eine (von statistischen Elementen freie) Theorie der Elementar-Teilchen sowie der Quanten-Phänomene enthält." Einstein on p. 15 in "Einheitliche Feldtheorie," unpublished manuscript, 6 July 1938, EA 2-121.)

Valentin Bargmann, mathematical physicist

Valentin Bargmann (1908–1989) was born in Berlin of Russian-Jewish parents. He began his studies at the local university in 1925 – he regularly saw Einstein in seminars, but would not approach him. His parents had experienced pogroms in Russia, and were certain they had to leave Germany in 1933. They went to Riga, while their son traveled to Zurich, where he continued his graduate studies. In 1936 Bargmann defended a thesis on, according to his own judgment, a "boring" problem of electron scattering in crystals, written under the guidance of Gregor Wentzel.[41] We have of course already seen some of his other early work: his 1934 study of spinors and semivectors, inspired by Wolfgang Pauli.

In 1937 Bargmann emigrated to the United States, as he believed there would be no jobs and no prospect of a permanent visa in Switzerland. That same year he was taken on by mathematician John von Neumann as a (non-salaried) assistant at the Institute for Advanced Study, and he joined Einstein and Bergmann in their study of the Kaluza–Klein theory.

In Bargmann's recollection, the working day began at ten o'clock when he and Peter Bergmann walked with Einstein to the Institute (see Figure 6.4). Discussions would last until lunchtime, when both again accompanied Einstein back to his house. They then all went to their own desks and worked out the issues debated. "Berg" and "Barg," as Einstein's secretary, Helen Dukas, had christened them, focused particularly on the calculations that needed to be done.

As in the case of semivectors, Wolfgang Pauli sent Einstein a letter that expressed his dislike of the latter's efforts. Bargmann was again the chosen channel of communication, although he now brought Einstein's ideas to Pauli, rather than the other way around. Pauli wrote in September 1938:

Herr Bargmann was recently here and reported on your work on the closed 5-dimensional continuum. But that is an old idea of *O. Klein*, and he has always emphasized the fact that the ψ-function gets a factor of e^{ix_5} and that changing the origin of x_5 corresponds to so called gauge group transformations. Besides our old, more fundamental differences of opinion, your Ansatz appears to me too special, because it is hardly necessary for the ψ-function to be a symmetric tensor.[42]

A symmetric second rank tensor represents a spin-2 field, yet electrons have a half-integer spin; Pauli thus pointed out that Einstein's charged sources, the higher

---

[41] See "Notes on V. Bargmann interview," 13 July 1984, interview by John Stachel and David Cassidy, EA 75-389. See also Bargmann (1979) and Halpern (2004), pp. 166–169.

[42] "Herr Bargmann [Valentin Bargmann had joined Bergmann and Einstein in their work on the Kaluza–Klein theory around this time] war kürzlich hier und hat von Ihrer Arbeit über das geschlossene 5-dimensionale Kontinuum berichtet. Das ist ja eine alte Idee von O. Klein, auch der Umstand, dass die ψ-Funktion der Wellenmechanik einen Faktor e^{ix_5} bekommt und dass die Änderung des Nullpunktes von x_5 der sog. Eichgruppe entspricht, wurde von ihm immer betont. Abgesehen von den alten mehr prinzipiellen Differenzen unserer Auffassungen, schien mir aber der von Ihnen gewählte Ansatz zu speziell zu sein, da wohl kaum die ψ-Funktion notwendig ein symmetrischer Tensor sein muss." Wolfgang Pauli to Einstein, 6 September 1938, EA 6-270; see also Doc. 527, pp. 598–599 in von Meyenn (1985), emphasis as in original.

The most important problem that was studied was how the field equations of the Einstein–Bergmann–Bargmann theory could give a charged particle solution, and in particular "the simplest solution of this kind." In trying to find this solution, Einstein did not work in a very structured way, according to Bargmann. He rather tried to retrieve the solution by trial and error. Einstein worked intensively on the problem, but Bargmann never saw Einstein "tensed or nervous."[43]

Bargmann did believe that Einstein had changed considerably in comparison with how he had seen him in Berlin: Einstein had become much less involved in public discussions and could scarcely be observed outside his office. Part of the reason lay in the fact that Einstein was famous and that this presented its difficulties; his office had to be moved once because people were peeking through the windows, and the few seminars that he gave were never publicly announced. Nevertheless, for a close associate like Bargmann, Einstein was easy to approach and he felt the latter to have been a fatherly friend who advised and assisted him in may ways.

Bargmann had already acquired a fairly strong intellectual independence before joining Bergmann and Einstein. He continued working with Einstein until 1943, when their second and last joint publication (on "bi-vectors") was submitted. According to Bargmann, Einstein said of his efforts in unified field theory that they were attempts to find the logically simplest theory in a given class of theories. Einstein then admitted that this statement in itself was too vague a characterization of a desired theory, yet that he was convinced it could be restated as a precise definition.[44] In 1943, Bargmann became an American citizen and began to undertake war related work, the study of shock waves, with John von Neumann.

order x^5-dependent metric fields, could hardly be considered a natural candidate for electrically charged matter. With the "more fundamental differences of opinion" he presumably referred to Einstein's familiar objections to quantum mechanics, and his long-standing refusal to dismiss classical field theory.

Pauli continued his letter by saying that in his opinion, Bargmann was more talented in mathematics than in physics: he would lack a good sense for how to put a physical problem in mathematical terms – exactly the quality that "should distinguish the theoretical physicist from the mathematician." His criticism however extended to Einstein too: "considering your current involvement with theoretical physics, this should hardly make a bad impression on you."[45]

[43] "[D]ie einfachste Lösung dieser Art"; "[G]ereizt oder nervös"; as in Bargmann (1979), p. 42.

[44] See Einstein and Bargmann (1944); see also Bargmann (1979), p. 44. The "bivector field" was again a concept that was "simple and natural enough to be of interest even apart from the physical problem we have in mind" (Einstein and Bargmann, 1944, p. 1).

[45] "[Bargmanns] Schwierigkeit war immer die, dass er kein richtiges Gefühl dafür hatte, wie man ein physikalisches Problem mathematisch anzusetzen hat – also eben das, was den Mathematiker vom theoretischen Physiker unterscheiden soll, fiel ihm schwer. Und in dieser Verbindung möchte ich nun mit einer kleinen Bosheit schliessen: Bei der Art Ihrer jetzigen Beschäftigung mit theoretischer Physik dürfte Ihnen das kaum sehr unangenehm auffallen." Wolfgang Pauli to Einstein, 6 September 1938, EA 6-270; see also Doc. 527, pp. 598–599 in von Meyenn (1985).

From 1941 on, Bargmann taught at Princeton University (apart from one term spent at the University of Pittsburgh), and he would carry the baton of mathematical physics at Princeton until his retirement. Bargmann's lectures on topics of his own research – whether they dealt with representations of the Lorentz group or second quantization – were known for their clarity and polish. His co-workers in Princeton found that his interests were broad, as his discussions on the foundations of physics with the philosopher Carl Hempel may exemplify. Bargmann's papers are not numerous, but his work has been quite influential, again according to his colleagues in mathematical physics. Shortly after receiving the Max Planck Medal of the German Physical Society, Valentin Bargmann died in 1989.[46]

Einstein let the criticism of Bargmann and himself pass and merely replied that Pauli "could easily be answered likewise." Perhaps Einstein intended with his remark to remind Pauli that he, too, had on occasion engaged himself with five-dimensional unification. In any case, Einstein did not agree that his work was superfluous: "the new work is only superficially similar to Klein's. It is simply a logical improvement of Kaluza's idea, that deserves to be taken seriously and examined accurately."[47]

Einstein did not elaborate further on how his theory would be different from Klein's but he was not deterred by Pauli's concerns. Well after their exchange, Einstein still spoke to Maurice Solovine of "a wonderful project that I am zealously elaborating with two young colleagues" – "a generalization of the theory of relativity of great logical simplicity."[48] He had told Peter Bergmann that he was quite pleased with the new five-dimensional approach and expressed "a lot of confidence in the theory as such."[49] Bergmann wrote in reply that he, too, had "great confidence in the theory because of its internal unity and the existence of 'wave type' fields."[50] Einstein and Bergmann were now *en route* to the greater goal: to re-derive typical quantum relations. "There is hope that in this way the [...] intolerable statistical foundation of physics can be overcome."[51]

---

[46] For more on Valentine Bargmann, see Lieb *et al.* (1976) and the interview that W. Aspray and A. Tucker conducted with Bargmann on 12 April 1984, available at the Princeton University website. See also J. R. Klauder, "Valentin Bargmann, April 6, 1908–July 20, 1989," *Bibliographical Memoirs of the National Academy of Sciences*, **76**, 1999, pp. 36–49, available online as well at the National Academy website.

[47] "Ihre Bosheit ist wohlbegründet und könnte leicht mit einer ähnlichen beantwortet werden. Die neue Arbeit hat aber mit der Kleins nur eine äusserliche Ähnlichkeit. Es ist einfach eine logische Verbesserung der Kaluzaschen Idee, die ernst genommen und genau geprüft zu werden verdient." Einstein to Wolfgang Pauli, 19 September 1939, EA 19-175; see also Doc. 530, on pp. 600–601 in von Meyenn (1985), some of Pauli's work on five dimensions is contained in Pauli (1933b).

[48] "[I]ch [bin] auf eine wunderbare Arbeit gestossen, an der ich mit zwei jungen Kollegen mit grossem Eifer arbeite. [...] Es ist eine Erweiterung der allgemeinen Relativitätstheorie von grosser logischer Einfachheit." Einstein to Maurice Solovine, 23 December 1938, on p. 76 in Einstein (1956).

[49] "[M]ein Vertrauen in die Theorie als solche ist sehr gross." Einstein to Bergmann, 5 August 1938, EA 6-271.

[50] "[A]uch ich [habe] grosses Zutrauen zur Theorie wegen ihrer inneren Geschlossenheit und der Existenz „wellenartiger" Feldvariabeln." Bergmann to Einstein, 15 August 1938, EA 6-272.

[51] "Es besteht Hoffnung, auf diese Weise die mir unerträgliche statistische Grundlage der Physik zu überwinden." Einstein to Maurice Solovine, 23 December 1938, p. 76 in Einstein (1956).

Figure 6.4. Princeton, 1940: Bergmann, Einstein and Bargmann (Lucien Aigner/ Corbis).

However, it is not straightforward to reconstruct from Einstein and Bergmann's correspondence which quantum laws they were aiming at. Bergmann's 1942 textbook on relativity mentions in its chapter on the five-dimensional theory that one of their goals had been to account for Heisenberg's uncertainty relations: a four-dimensional description of a five-dimensional world would necessarily be incomplete and indeterminate.[52] Yet, Heisenberg's relations are not directly addressed in Einstein and Bergmann's contemporary correspondence, nor in their papers; they likely remained only a distant objective.

At certain points in their letters there was however mention of a "de Broglie frequency." Yet, it seems that the relations which contain a matter wave were introduced initially only to serve the purpose of checking the consistency of the theory, for example, to see whether the equations allow solutions with negative masses. The "wave type" field that Bergmann referred to in his letter to Einstein quoted above corresponded to a metric that has a wave propagating in the fifth direction,

---

[52] See Bergmann (1942), p. 272; see also Halpern (2004), p. 173.

the momentum of which would give an electrically charged source term in the field equations.[53]

For Einstein, a plausible first step towards a re-derivation of any quantum law would be the construction of a non-singular charged particle solution. Just after having written the paper with Bergmann, published in the *Annals of Mathematics* in July 1938, he started his attempts at realizing such a non-singular object. "I hope," Einstein told Bergmann, "that this will now go smoothly, without the eternal goddess once more sticking out her tongue at us."[54] Of course he had not forgotten the pitfalls that such a quest had presented him in the past.

The search for a particle solution lasted for some three years.[55] One equation under consideration was the following:

Bergmann to Einstein, August 1938:

[...] in our case the lowest power of the equation for the current is:

$$+\frac{2\pi}{\lambda}\omega[\alpha(3\alpha+2\beta)] = \alpha_2\left(\phi'' + \frac{2}{\rho}\phi'\right) \tag{6.18}$$

i.e. the relative sign of the charge (term on the right hand side) and ω/λ also depends on the sign of the square bracket which is different for both kinds of particles.[56]

Here λ is the scale of the fifth direction, ω the frequency of a wave traveling in the fifth direction, α_2 is a coupling constant, ρ is a radial coordinate ($\rho^2 = r^2 + a^2$ with a a constant and r the usual radial coordinate) and ϕ is the electrostatic potential which is differentiated with respect to ρ. The equation under consideration was quite likely the 0-component of the Maxwell equations (6.17), i.e. the electrostatic Poisson equation, written in the leading terms at large distances. In this interpretation the α and β terms stem from the four-dimensional curved metric. The solution of the full equation would correspond to an object represented by a wave propagating in the fifth direction and curving the four-dimensional spacetime; dimensional reduction from five to four dimensions would yield a massive body sitting somewhere in 4-space, yet carrying a momentum in the fifth direction, which would give the electric charge.

---

[53] "We had put $\gamma_{\mu\nu} = \psi_{\mu\nu}e^{2i\pi x^5/\lambda} + \bar\psi_{\mu\nu}e^{-2i\pi x^5/\lambda}$ (= real quantity, bar means complex conjugate) [...] We had further taken for our stationary problem $\psi_{\mu\nu} = \chi_{\mu\nu}e^{i\omega t}$, $\chi_{\mu\nu}$ = function of the spatial coordinates alone." ("Unsere Setzung war dann $\gamma_{\mu\nu} = \psi_{\mu\nu}e^{2i\pi x^5/\lambda} + \bar\psi_{\mu\nu}e^{-2i\pi x^5/\lambda}$ (= reelle Grösse, Querstrich bedeutet konjugiert komplex.) [...] Weiter hatten wir für unser zeitlich stationäres Problem gesetzt: $\psi_{\mu\nu} = \chi_{\mu\nu}e^{i\omega t}$, $\chi_{\mu\nu}$ = Funktion der räumlichen Koordinaten allein.") Peter Bergmann to Einstein, 4 August 1938, EA 6-270.
[54] "Ich hoffe, dass dies nun glatt gehen wird, ohne dass uns die ewige Göttin erneut die Zunge herausstrecken wird." Einstein to Peter Bergmann, 4 July 1938, EA 6-253.
[55] According to Bargmann (1979), pp. 42–43.
[56] "Für unseren Fall heisst die niederste Potenz der Stromgleichung: $+\frac{2\pi}{\lambda}\omega[\alpha(3\alpha+2\beta)] = \alpha_2(\phi'' + \frac{2}{\rho}\phi')$. Das heisst, das gegenseitige Vorzeichen der Ladung (Glied rechts) und von ω/λ hängt noch vom Vorzeichen der eckigen Klammer ab, und dieses ist für beide Teilchensorten verschieden." Peter Bergmann to Einstein, 4 August 1938, EA 6-270.

What type of solution they hoped to retrieve can be read in the following note from Einstein to Bergmann:

The treatment of our centrally symmetric problem is still mysterious to me. [...] Maybe we will have to introduce ρ again. But I still do not have a comprehensive view of the situation, particularly as ϕ at ∞ has to go as ϵ/r. One would have to know a lot about differential equations to find this solution. Perhaps Bargmann will find a way. It is certainly not easy. In any case, one first needs to find a description at large r and then strive for a convergent description at the origin.[57]

So Einstein wanted a particular solution that drops off as ϵ/r at large distances: far away from the object the field should look like that of an electric monopole. He also wanted a non-singular, convergent series for the potential at the location of the source. (At some point, he expected that substituting ρ instead of r would give a convergent series.) Thus, by solving differential equations like (6.18), Einstein was trying to find a non-singular charged particle solution to his Kaluza–Klein field equations.[58]

But how might he have imagined that such a solution could yield any quantum relations? We cannot answer this question with any strong certainty. The only lead that we have identified is that on a number of occasions, the frequency of charged wave-solutions was indeed identified as a "de Broglie frequency."[59] One could then allow the relation $E = h\nu$ for these solutions, and from the de Broglie relation it would follow that such five dimensional photons carry a charge given by $p_5 = h/\lambda$. Following Oskar Klein's logic would now suggest the observation that the quantized nature of the electric charge is in accordance with the compactification of the fifth direction. Yet, as we saw earlier, Einstein never assumed a discrete charge spectrum, so despite the fact that at times he thought of the matter waves as de Broglie waves, we do not think that he intended to follow Klein's dual reasoning – for one thing, he never mentioned Klein's argument in any of his papers. Furthermore, as we have seen, consistency would require that in that case one would have to make the transition from classical fields to quantum fields, i.e. from numbers to operators: without quantization, the momentum in the fifth direction of the wave-field would still have a continuous range of values; see (6.14). Einstein however was known to think of second quantization as sinning squared.[60]

[57] "Die Behandlung unseres zentralsymmetrischen Problems ist mir immer noch dunkel. [...] Vielleicht muss man wieder ρ einführen. Aber ich kann die Sache noch nicht übersehen, zumal ϕ im ∞ sich wie ϵ/r verhalten muss. Um dies zu finden, sollte man viel von Differentialgleichungen wissen. Vielleicht findet Bargmann einen Weg. Einfach ist es keinesfalls. Jedenfalls muss man zunächst eine Grenzdarstellung für grosse r suchen und dann erst eine im Nullpunkt konvergierende Darstellung anstreben." Einstein to Peter Bergmann, 5 August 1938, EA 6-271.

[58] Compare also with Bargmann (1979), pp. 42–43.

[59] As in for example Einstein to Peter Bergmann, 21 June 1938, EA 6-249.

[60] According to a 1963 interview with Oskar Klein by Thomas Kuhn and John Heilbron for the Archive for History of Quantum Physics, cited on p. 173 in Halpern (2004).

We feel that instead of following Klein's argument, Einstein would rather have hoped that the generalized Poisson equation (6.18), with the boundary conditions that he imposed and combined with the generalized Einstein equations (6.16), would produce a discrete spectrum of charged solutions. That is to say, the charges p_5 would have a discrete spectrum because the field equations and boundary conditions would restrict the integral over the charged metric fields to a discrete set of values. This would actually have provided an alternative *derivation*, not only of the quantum of charge, but also of the de Broglie relation: because of the compactification, there would then be a fundamental constant in the theory with the dimension of angular momentum

$$p_5 \lambda = h. \qquad (6.19)$$

One could set this to the value of Planck's constant and arrive at a value for λ. Such a scenario would avoid the introduction of some quantization procedure that Einstein would have regarded as epistemologically ad hoc and thus unnatural. Instead, the discretely charged solutions of the classical field equations, together with the compactification radius, would yield Planck's constant: quantum concepts, such as h, would then not play a "bottom-up" role in the theory, but be properly deduced.

We have to admit, however, that there is no explicit mention either in Einstein's correspondence that we have seen, nor in his articles, that the discrete charge spectrum was the key result that he was looking for in his Kaluza–Klein theory – nor that the above argument would be his method for re-deriving h.[61] We do know for certain that he wanted to find a non-singular particle solution; how this should reproduce quantum relations or properties of elementary particles will regrettably remain guesswork. Nevertheless, given Einstein's silence on Klein's treatment of the compactification, combined with his search for a particle solution, we would argue that the above "derivation" of Planck's quantum and de Broglie wavelength relation would have appealed more to him than Klein's straightforward invocation of this quantum law.

Einstein's search for a particle solution was unsuccessful, as was reported in the second and last paper on the theory. The article came out in 1941 but was quite likely written sooner, as its slightly different approach – grounding the theory on

---

[61] The following quote is however quite suggestive of a wish on Einstein's part to retrieve a discrete charge spectrum from a theory with a fundamental length scale, and thereby find a way to an alternative account of quantum relations: "If one does *not* want to introduce rods and clocks into the theory, then one must have a structural theory in which a fundamental length enters, which then leads to a solution in which this length occurs, so that there no longer exists a continuous sequence of "similar" solutions. This is indeed the case in the present quantum theory, but has nothing to do with its basic characteristics. Any theory which has a universal length in its foundations and, on the basis of this circumstance, qualitatively distinguished solutions of definite extent would offer the same thing with respect to the question envisioned here." Einstein to W. F. G. Swann, 24 January 1942, EA 20-624, as in Stachel (1986), on p. 394 in Stachel (2002).

five-dimensional general covariance – is already mentioned in a letter from Einstein to Bargmann from 1939. In any case, the failure to find a particle solution was the reason why Einstein again abandoned the Kaluza–Klein theory.[62]

Einstein returned to five dimensions on one last occasion, in an article that appeared in the *Annals of Mathematics* in 1943. It was written together with Wolfgang Pauli and contained a proof that non-singular particle solutions are not possible in the Kaluza theory, or any theory of more than four dimensions.[63]

It is quite striking that Pauli was Einstein's co-author. We have seen Pauli repeatedly and strongly criticize Einstein's efforts since the very inception of the unified field theory program; on all of those occasions, Pauli's objective appears to have been to convince Einstein of the futility of his attempts. He had followed Einstein's work quite closely over the years, touching upon similar themes in some of his own research. On occasion he even published follow-up articles,[64] although Pauli himself was interested primarily in the elaboration of quantum theories.

The 1943 article was an extension of a proof that Einstein had given on the non-existence of non-singular particle solutions in four-dimensional general relativity,[65] and it was written while Pauli stayed in Princeton during World War II. We believe that one should think of Pauli's involvement in this project as one more attempt to convince Einstein that he was pursuing a dead end – as one more attempt to win Einstein over for the quantum theory program. When viewing their article from the perspective of their past exchanges on unified field theories, one easily imagines that Pauli pursued or proposed the topic to demonstrate once and for all that Einstein entertained a vain hope by his search for particle solutions. The conclusion of their article was in any case nothing short of a no-go theorem for non-singular Kaluza theory particles. Ironically, however, soliton solutions were in fact found in the Kaluza theory with a compact fifth dimension much later, in 1983.[66]

Pauli kept following Einstein's work after his return to Europe, and also kept expressing his criticism. In 1946, only a few years after their joint paper, Pauli told Einstein, "as before," that in his opinion, "not in the least because of the negative results of your own numerous attempts, classical field theory in any form is a completely squeezed out lemon that can impossibly yield anything new!"[67]

---

[62] According to Bargmann's recollections, see Bargmann (1979), p. 43; Einstein's letter was Einstein to Valentin Bargmann, 9 July 1939, EA 6-207.

[63] See Einstein and Pauli (1943).

[64] See for example Pauli and Solomon (1932).

[65] See Einstein (1941).

[66] On these solitons, see Sorkin (1983), Gross and Perry (1983). For analysis that compares the Einstein–Pauli proof with the existence of these solutions, see van Dongen (2002), pp. 197–206, and Giulini (2008), pp. 63–65.

[67] "Meine persönliche Ueberzeugung ist nach wie vor – nicht zuletzt infolge der negativen Ergebnisse Ihrer eigenen zahlreichen Versuche – dass die klassische Feldtheorie in jeder Form eine völlig ausgepresste Zitrone ist, aus der unmöglich noch etwas neues herauskommen kann!" Wolfgang Pauli to Einstein, 19 September 1946, EA 19-182.

Einstein's next letter to Pauli announced new confidence in yet another unified field theory; this one had a "Hermitian" metric. He was however having some difficulty in finding non-singular solutions to its field equations.[68]

6.4 Conclusion: classical field theory and quantization

Einstein wanted the quantum relations to follow, not to guide. As we saw, in all of his work on Kaluza–Klein theory, Einstein consistently avoided the point made by Klein – that through the quantum de Broglie relation, the compactification of the fifth direction could imply a discrete charge spectrum. Einstein did not want to introduce $\hbar$ from the outset, rather, he likely wanted to deduce its existence from a unified and mathematically natural set of field equations. Once a satisfactory set of field laws was found in the five-dimensional theory, he immediately tried to find a non-singular particle solution; presumably, such solutions could express nature's atomicity and yield the empirical relations that quantum theory indeed contained, but had failed to explain.

Einstein did not change his perspective after he had left all the varieties of the Kaluza–Klein theory. His assistant Bergmann, however, later worked on quantum gravity, on a theory of gravity in which the gravitational interaction would be quantized in advance. In 1949, just as Einstein was recovering from a laparotomy, Bergmann, now at Syracuse University, wrote to Einstein and asked if they could have a discussion sometime:

> As anyone can only be a crank about his own ideas, and as you are someone who combines perseverance with the ability to acknowledge that his hypothesis could go wrong (usually one can only find just one of these qualities, mostly the latter) I would appreciate very much simply talking to you and hearing your comments; whether we appreciate the same or not, what we want is sufficiently related that we can easily communicate.[69]

Einstein, however, discouraged Bergmann from traveling to Princeton. Not only was he recovering from surgery, he also could not see the immediate purpose of further discussion between them:

> You are looking for an independent and new way to solve the fundamental problems. In this endeavor no one can help you, least of all someone who has somewhat fixed ideas. For instance, you know that on the basis of certain considerations I am convinced that the probability concept should not be primarily included in the description of reality, while you

---

[68] Einstein to Wolfgang Pauli, 1 April 1948, EA 19-184.

[69] "Da nun jeder Mensch nur seinen eigenen Erfindungen gegenueber crank ist, und da Sie sogar ein Mensch sind, der Beharrlichkeit verbindet mit dem Zugestaendnis, seine Hypothese koennte schiefgehen (im allgemeinen kann ich immer nur die eine dieser Charaktereigenschaften an einem sehen, meistens die letztere), so wuerde mir sehr viel daran liegen, einfach Ihnen zu erzaehlen und Ihre Bemerkungen zu hoeren; moegen wir dieselbe Bewertung haben oder nicht, was wir wollen, ist genuegend verwandt, dass wir uns leicht verstaendigen koennen." Peter Bergmann to Einstein, 24 January 1949, EA 6-282.

seem to believe that one should first formulate a field theory and subsequently "quantize" it. This is in keeping with the view of most contemporaries.[70]

Quantization, according to Einstein, was mathematically an ad hoc procedure. It implied one would superimpose the quantum laws right at the beginning of the formulation of a theory; in the language of the Solovine schema, it meant that the quanta were included in a fundamental set of axioms A, whereas their true place was lower, on the level of the phenomena E; quantization would be some easy, inductive, bottom-up step. Likewise, arguments that assumed the de Broglie relation went the wrong way: they entailed that one could start out with observed phenomena and phenomenological, physical relations in theory construction. But this was not in keeping with the desire to come to true and deep understanding of the quantum: to really understand the atomistic character of nature, to truly fathom the occurrence of quantum probabilities, the Heisenberg relations or the values of nature's fundamental constants, these all had to follow from a mathematically natural and unifying set of axioms, as Mercury's perihelion motion and the bending of light-rays had followed from the field equations derived from the Riemann tensor. Comparing the work on Kaluza–Klein theory with the search for general relativity, Einstein seems to have wanted to follow a top-down mathematical route, instead of a dual approach.

The Einstein–Bergmann–Bargmann version of Kaluza–Klein theory provided a mathematically natural and logically simple set of axioms.[71] The entire structure fits our Solovine schema-like figure (Figure 3.1) rather well, with the generally covariant theory of relativity on the bottom left and the gauge invariant Maxwell theory on the right. Kaluza had made the natural and intuitive leap to five dimensions and Einstein, Bergmann and Bargmann had tried to write the most general field

---

[70] "Sie suchen einen selbständigen und neuen Weg zur Lösung der prinzipiellen Schwierigkeiten. Bei diesem Bestreben kann einem Niemand helfen, am wenigsten Einer, der einigermassen fixierte Ideen hat. Sie wissen z. B., dass ich auf Grund gewisser Überlegungen fest glaube, dass der Wahrscheinlichkeitsbegriff nicht primär in die Reali[t]ätsbeschreibung eingehen darf, während Sie daran zu glauben scheinen, dass man zuerst eine Feldtheorie aufzustellen und diese dann nachträglich zu "quantisieren" hat. Dies stimmt ja mit der Ansicht der meisten Zeitgenossen überein." Einstein to Peter Bergmann, 26 January 1949, EA 6-283. Bergmann had proposed an approach in which a "skeletonized" general version of Einstein's gravity theory – generally covariant but not based on Riemannian geometry and metric tensors – would in the end go through a quantization procedure; see his first paper on the subject (Bergmann, 1949, pp. 680–681). Bergmann's proposals were the first steps in what would become known as the "canonical quantum gravity program," see also Bergmann and Brunings (1949), Bergmann *et al.* (1950), Bergmann (1992). Einstein did not agree to the approach: "Your effort to abstractly carry through a field theory without having at your disposal the formal nature of the field quantities in advance does not seem favorable to me, for it is formally too poor and vague." ("Ihr Versuch, eine Feldtheorie abstrakt durchzuführen, ohne von vornherein über die formale Natur der Feldgrössen zu verfügen, erscheint mir nicht glücklich, weil formal zu arm und unbestimmt.") Einstein to Peter Bergmann, 26 January 1949, EA 6-283.

[71] Compare with "natural axioms" ("[N]atürliche Axiome"), in "Verallgemeinerte Kaluza-Theorie der Elektrizität," single page draft manuscript, EA 75-367, date 1938 or 1939; "great logical simplicity" ("grosser logischer Einfachheit"), Einstein to Maurice Solovine, 23 December 1938, on p. 76 in Einstein (1956).

equations possible under the assumption of periodicity. The requirement of a four-dimensional limit implied that the extra dimension needed to be small. How small, Einstein and his assistants did not determine: Einstein might have expected that result to follow once they had deduced a non-singular particle solution of the field equations. Just as in Figure 3.1 the S_7 follow from the unified A', and explain and describe additional elements from experience E, the particle solutions of the Kaluza–Klein theory were expected to elucidate the quantum laws and fundamental constants.

Einstein could not find suitable solutions in five dimensions, however, and moved away from the theory. Yet he did not give up on his conviction that the deduction of such solutions should unravel the quantum: he maintained that position until the end of his life. In the 1955 edition of *The Meaning of Relativity*, a book that contained a reworking of old lectures and to which appendices on various unified field theories had continuously been added, he wrote:

Is it conceivable that a field theory permits one to understand the atomistic and quantum structure of reality? Almost everybody will answer this question with "no." But I believe that at the present time nobody knows anything reliable about it. This is so because we cannot judge in what manner and how strongly the exclusion of singularities reduces the manifold of solutions. We do not possess any method at all to derive systematically solutions that are free of singularities. Approximation methods are of no avail since one never knows whether or not there exists to a particular approximate solution an exact solution *free of singularities*. For this reason we cannot at present compare the content of a nonlinear field theory with experience.[72]

---

[72] As in Einstein (1955), p. 165, emphasis and text as in Einstein's "Appendix II," written in December 1954.

7

The method and the quantum

To complete the picture of Einstein's later thought, we end by turning briefly to the critique of quantum theory that he formulated in the same years as he was elaborating unified field theories. We intend to address this critique in light of the historical and methodological perspective that we have taken with regard to the unified field theory work. In this way, we hope to come to a more complete understanding of Einstein's position.

In short, one can say that Einstein's attitude to quantum theory was conditioned by a conviction that the theory was at best a mere phenomenological description of reality. We argue that this attitude was in part due to the methods employed in formulating quantum theory, as these deviated from his gradually developing unified field theory methodology. In addition, it found its expression in the epistemological objections that Einstein formulated, such as the argument contained in the EPR article.

The chapter is organized as follows. We first give a characterization of quantum research in terms of its methods and practices before quantum theory had reached its more complete formulations. The continuously evolving description of quantum phenomena prior to 1925 is generally referred to as the "old quantum theory,"[1] and we will be touching on Einstein's role in its development. We then briefly address the emergence of matrix and wave mechanics and discuss Einstein's first reactions to these theories. Finally, we intend to place his dismissive attitude towards the established quantum theory in the context of his attempts at elaborating unified field theories. On the whole, we intend to focus on the practices of the quantum program and compare these to Einstein's ideals and attempts.

---

[1] On the term "old quantum theory," see for example p. 94 in Jammer (1989). Jammer included in the discussion the Bohr–Sommerfeld quantization rules, the correspondence principle, Alfred Landé's account of the anomalous Zeeman effect and Pauli's introduction of the exclusion principle.

7.1 The old quantum theory

In its early years, the quantum program roughly consisted of two strands: on the one hand there were studies in radiation theory, beginning in 1900 with Planck's quantum hypothesis. On the other hand we find studies of atomic structure. These helped shape the quantum program in particular through the 1913 Bohr model of the atom. Einstein's most important contributions to the old quantum theory were mostly contributions to the first subject, as they centered foremost on the light quantum – the light quantum remained the focus of his attention even if, for shorter periods, he also worked on problems like the specific heat of solids or the counting of states of a gas of quantum particles. Most of the scholars who eventually shaped the full theory of quantum mechanics – men like Niels Bohr and Arnold Sommerfeld, and the latter's many highly successful students, such as Werner Heisenberg and Wolfgang Pauli – were however mostly involved in the second strand, the problem of atomic structure. This suggests that we might learn more about Einstein's later apartness when we compare the practices in theory construction that were prevalent in this group to Einstein's methodological beliefs.

Generally speaking, one could say that the quantum program initially introduced principles and rules that were often in opposition to "classical" physics,[2] yet depended on classical physics to be formulated. The resulting structure, i.e. the old quantum theory, would not easily be reconcilable with an ideal of logical simplicity. The Bohr–Sommerfeld quantization rules, for example, were at odds with any understanding of classical physics, yet were couched in classical language. Variance with a maxim of logical simplicity was also exhibited in Bohr's earliest work on his atom model, just as observation and experiment shaped this model more than their role in the older Einstein's epistemological ideals would have licensed. "The whole thing was immediately clear to me," Bohr said, "as soon as I saw Balmer's formula."[3] Bohr employed the empirical Balmer formula for frequency regularities observed in the spectral lines of hydrogen to fix his atomic model. He also derived the Balmer spectrum retroactively from his model. Model fixing occurred as the Balmer formula suggested divorcing the optical frequencies from the electron's mechanical frequency: the difference between the terms in the Balmer expression,

$$v = Rc \left(\frac{1}{\tau_1^2} - \frac{1}{\tau_2^2} \right), \tag{7.1}$$

---

[2] One should be careful introducing "classical physics" as an identifier when speaking of early twentieth century physical theory; different historical actors would have understood different things by the term. Max Planck in 1911 at the first Solvay conference used it in a way that modern readers would recognize and we will invoke the same usage here; Planck spoke of classical theory as, largely, mechanics and electrodynamics, extended with the kinetic theory of gases and relativity; this body of knowledge had become too narrow in his opinion, as illustrated by its contradictions with the black body distribution law. See Staley (2005), in particular p. 555 for Planck's comments.

[3] Niels Bohr, as cited in Heilbron and Kuhn (1969), p. 265.

suggested that emission occurred in transitions between quantized states of the atom (R and c represent the Rydberg constant and speed of light, respectively, and τ_1 and τ_2 are integers, which according to Bohr denoted quantum states; ν is the frequency of the observed light).

Empirical model fixing however also took place in a more ad hoc way. For instance, an undetermined parameter in Bohr's expression for the energy levels of the atom had been put to a value of exactly $1/2$, only to match the observed spectral series (this gave $\tau h\omega/2$, with τ an integer, h Planck's constant and ω the electron's orbital frequency).[4] Finally, Bohr's conceptual interpretation of his model was equivocal, as was exhibited by the two different derivations of the Balmer spectrum included in his 1913 paper: in one derivation, a free electron that was being captured into a stationary state emitted τ energy quanta with energy $\omega/2$; in another it emitted a single quantum of energy $\tau\omega/2$. To many, Bohr's theory indeed only remained an "ingenious play with numbers and formulas."[5]

To be sure, Einstein was reportedly quite excited by Bohr's model when he first heard of it, and it came out before the final formulation of general relativity had been achieved and his epistemological transition had run its course. John Stachel has even entertained the idea that Einstein had earlier developed ideas himself that were broadly similar to Bohr's. Yet, Stachel has also suggested, and we agree with his suggestion, that Einstein's approach to physics would likely have led him away from Bohr's problem[6] – the above characterization of Bohr's first article in any case does show a disaccord with Einstein's later methodological ideas.

Atomic theory continued to develop in ways that were increasingly at variance with the direction in which Einstein's epistemology and practice would evolve. Initially, theoretical atomic model building received a boost through Arnold Sommerfeld's generalization of the Bohr quantization to more degrees of freedom. Sommerfeld held that these new conditions were "unproved, and perhaps incapable of being proved."[7] Yet, in combination with a relativistic treatment of the orbiting electron, he could deduce the fine structure of the hydrogen spectrum. Progress along these lines however came to a halt by 1919: multiplet and anomalous Zeeman spectra proved elusive to a treatment through the quantum conditions, and Sommerfeld began to take a different approach. Instead of starting out with a definite atomic model to which classical mechanics and the quantization rules were to be applied, with the derived results to be compared with observed spectral

---

[4] Bohr (1913), on p. 136 in ter Haar (1967), actually argued that the factor was due to averaging over the electron's energy before and after its capture by an atom. However, historians are agreed that the factor was more likely fixed by comparison with the Balmer series, and they have qualified the averaging argument as an ad hoc rationalization introduced a posteriori; see Heilbron and Kuhn (1969), pp. 271–272.

[5] According to historians John Heilbron and Thomas Kuhn (1969), p. 266; for the observation of Bohr's equivocalness, see their p. 274.

[6] See Stachel (1986), pp. 368–371.

[7] Sommerfeld as quoted in Jammer (1989), p. 99; "unbewiesene und [...] vielleicht unbeweisbare Grundlegung der Quantentheorie," p. 6 in Sommerfeld (1916).

lines, he now chose to begin with the spectral lines, then try to retrieve from these inductively the atomic energy levels, and associate with the latter certain quantum numbers. The focus was on experimental regularities; the construction of explicit models was to be postponed.[8]

The non-appearance in multiplets of certain expected spectral lines suggested the absence of particular atomic transitions. Unearthing the selection rule at work might thus reveal something about the internal structure of the atom; Sommerfeld proposed that the selection rule at hand applied to a novel "internal quantum number." The latter's geometrical or conceptual interpretation, however, was entirely unclear. Sommerfeld further qualified the form of his proposed selection principle – the internal quantum number could only change by $+1$, -1 or 0 – as "somewhat arbitrary" and the choice for inner quantum numbers as "not definitive."[9] The introduction of the internal quantum number illustrates that experiment started leading the way in Sommerfeld's studies. Historians indeed have found that the inner quantum number was put forward merely as a numbering device and that Sommerfeld, in his approach, tried to find clues of an underlying model in essentially an "ad hoc" and "phenomenological" fashion.[10]

Sommerfeld briefly introduced Einstein in October 1921 to his ideas about the internal quantum number, more than one and a half years after he had submitted his major article on the subject;[11] clearly, he did not expect Einstein to have kept abreast with the confusing developments in studies of atomic structure. He put their approaches to quantum problems in relative perspective: "you ponder the fundamental questions concerning the light quantum. Because I do not feel the strength in me for that, I content myself with looking into the particulars of the quantum magic in the spectra."[12] Sommerfeld realized that his proposals were often rather eclectic, and, in a later letter written to his wife, he again observed a marked difference with Einstein: "I was always content if I could explain mathematically a certain complex of phenomena, without worrying too much that there are other things that do not fit in. Einstein, who always looks at the whole, makes his life

[8] For Sommerfeld's reversal, see Forman (1970), p. 186, Cassidy (2008), pp. 106–107, Seth (2008), pp. 340–341.
[9] On the conceptual interpretation, see Sommerfeld (1920), p. 231; "einigermassen willkürlich" [...] "nicht bindend"; p. 234 in Sommerfeld (1920). Sommerfeld also told Einstein that he "could not imagine anything" when thinking about the internal quantum numbers ("[D]ie „inneren Quantenzahlen" [...] unter denen ich mir aber gar nichts denken kann.") 17 October 1921, Doc. 274 in Buchwald *et al.* (2009).
[10] See Jammer (1989), pp. 125–127. See also Forman (1970), p. 186, Cassidy (2008), p. 107. Suman Seth (2008), p. 347, however disagrees with the "ad hoc" qualification, emphasizing instead that Sommerfeld was led by "a delicate sense of aesthetics." Yet, Sommerfeld's aesthetic lyricisms seem to be just that, rather than appearing to have directed him in his work; internal consistency, for example, was not of immediate concern to Sommerfeld.
[11] Sommerfeld to Einstein, 17 October 1921, Doc. 274 in Buchwald *et al.* (2009); the article is Sommerfeld (1920).
[12] "Sie grübeln mit Ihren Gedanken an den grundsätzlichen Fragen der Lichtquanten herum. Ich begnüge mich damit, weil ich zu jenem nicht die Kraft in mir fühle, den Einzelheiten des Quantenzaubers in den Spektren nachzugehen." Arnold Sommerfeld to Einstein, 17 October 1921, Doc. 274 in Buchwald *et al.* (2009).

more difficult."[13] Sommerfeld himself would quite likely have agreed that his phenomenological and eclectic approach was different from Einstein's unified field theory methodology.

Eclecticism also surfaced in studies of the anomalous Zeeman effect, that is, the anomalous splitting of spectral lines in a magnetic field; the Zeeman splitting occurred in an anomalous fashion whenever spectral lines exhibited a multiplet structure. Understanding the effect promised an understanding of the multiplets (and vice versa), and eventually, thus, an understanding of atomic structure. In 1921 Alfred Landé, another former student of Sommerfeld, arrived on purely empirical grounds at a set of rules that captured the complicated lines of the Zeeman effect in terms of three quantum numbers, j, k and m, and an empirically established gyromagnetic factor g; j was Sommerfeld's original internal quantum number and was now interpreted as the atom's total angular momentum, k was the valence electron's angular momentum and m the magnetic quantum number. Landé's recipe for Zeeman terms was however not consistent with any theoretical ideas: he had controversially introduced half-integer values for m, and the underlying dynamics that should produce the Zeeman splitting remained obscure.[14]

The eventual model interpretation that was proposed by Werner Heisenberg entailed that a valence electron would share a half unit of its angular momentum with the atom's core, i.e. the nucleus plus the lower lying electrons. This "core model" violated classical and quantum principles, but Heisenberg tried to dispel objections by stating that "success sanctifies the means."[15] Sommerfeld once again informed Einstein of these developments and added that "everything works out but yet in the deepest remains unclear. I can only advance the craft of the quantum, you have to make its philosophy."[16] With his last remark Sommerfeld intended to encourage Einstein to pursue further his latest thought on the light quantum (the letter was written during the excitement over the 1921 fallacious wave-particle experiment conducted by Geiger and Bothe mentioned earlier) – but, likely, he also intended to woo Einstein into studies of quantum numbers and spectra. Einstein, however, remained at a distance from the subject.

Wolfgang Pauli complained to Bohr early in 1924 that quantum theorists had come to a situation in which they were inconsistently using integer and half-integer

[13] "Ich war immer zufrieden, wenn ich einen gewissen Complex von Tatsachen mathematisch erklären konnte, ohne mich zu sehr zu beunruhigen, dass es andere Dinge gibt, die nicht herein passen. Einstein, der immer auf's Ganze blickt, macht sich das Leben schwerer." Arnold Sommerfeld to Johanna Sommerfeld, 24 December 1928, cited on p. 23 in Eckert and Märker (2004).

[14] On Landé's work, see in particular Forman (1970); see also Jammer (1989), pp. 126–130. Cassidy (2008), pp. 111–112.

[15] Heisenberg, as in Cassidy (2008), p. 109.

[16] "Alles klappt, bleibt aber doch im tiefsten Grunde unklar. Ich kann nur die Technik der Quanten fördern, Sie müssen ihre Philosophie machen." Sommerfeld to Einstein, 11 January 1922, pp. 95–97 in Einstein and Sommerfeld (1968), on p. 97; translation as in Seth (2008), p. 335.

quantum numbers as they pleased, with only the objective of finding agreement with experiment in mind.[17] Eventually Pauli himself found a way to interpret the role attributed to the angular momentum of the atom's core as in reality due to a "classically not describable, two-fold ambiguity"[18] of the optical electron. This resolved some observational difficulties – for instance, in the old models the Zeeman splitting needed to depend on the atomic number, which observation contradicted – but many conceptual problems remained. He also formulated the exclusion principle, which was again poorly motivated from the theoretical point of view, but did fit the phenomena rather well. As Pauli put it: "We cannot give a further justification for this rule, but it seems to be a very faithful one."[19]

The exclusion principle was typical in this respect of many of the ideas of the old quantum theory. One partly succeeded in reproducing theoretically the observational data, yet strict and consistent proofs often could not be given – in fact, the descriptions of the old theory often contained conflicting elements. The primary justification of such attempts was that they somehow managed to catalog, order, "save" the phenomena. This reflected the empirically driven methodology of the Sommerfeld school approach, as well as the empiricist practices of Bohr: Bohr found that his atomic models had been "guessed on the basis of the available data, and not obtained from any theoretical calculations."[20] He did not aim at removing the logical complexities, but rather at an ever greater exacerbation of them, such that these "mysteries [...] with every step forward [...] become ever more pronounced."[21]

Pauli, in his 1945 Nobel acceptance speech,[22] discussed how the approaches of both the Bohr and the Sommerfeld schools had largely directed the developments in quantum theory in the early 1920s. Most, if not all quantum theorists would pass through Copenhagen, and, as pointed out before, many had been educated in Sommerfeld's Munich, Pauli being a case in point. Historian Suman Seth has recently identified an interesting difference between these two schools. Sommerfeld, according to Seth, largely focused on solving particular problems, and avoided questions of principle. Bohr, and of course Einstein too, were instead drawn to foundational issues; they were rather engaged in a "physics of principles."[23] Einstein himself seemed to have sensed the same distinction:

There are those that vex principles ["Prinzipienfuchser"] and virtuosi. All three of us [i.e. Einstein, Bohr and Paul Ehrenfest, to whom Einstein was writing] belong to the first kind

17 See Cassidy (2008), p. 111.
18 "[K]lassisch nicht beschreibbare Zweideutigkeit," (Pauli, 1925, p. 765).
19 "Eine nähere Begründung für diese Regel können wir nicht geben, sie scheint sich jedoch von selbst als sehr naturgemäss darzubieten." As in Pauli (1925), p. 776. For more on the history of the exclusion principle, see Jammer (1989), pp. 130–150, Massimi (2005).
20 "[...] aus Erfahrungen erschlossen, oder, wenn Sie wollen, erraten, nicht aus irgendwelchen theoretischen Berechnungen gewonnen." Bohr as quoted by Werner Heisenberg in "Der Begriff „Verstehen" in der modernen Physik," on p. 63 in Heisenberg (1969).
21 "„Rätsel" [...] bei jedem Fortschritte [...] in ein immer schärferes Licht zu treten," p. VI in Bohr (1922).
22 Cited in Seth (2008), p. 337.
23 See Seth (2007), p. 41. I am also grateful to Suman Seth for sharing with me a chapter of his forthcoming book.

(at least the two of us) and have little virtuosic talent. So effect at encounter with distinct virtuosi (Born or Debye): discouragement. By the way, it works similarly the other way around.[24]

Although Peter Debye was yet another Sommerfeld student, Einstein did not explicitly mention Sommerfeld himself. Yet, it is clear that the latter's work on atomic spectra and his disinclination to study foundational issues would easily qualify him as another "virtuoso." Sommerfeld's problem oriented approach did not clearly resonate with Einstein's principle oriented way of doing physics.

According to Einstein, Bohr also aimed at a physics of principles. Yet, obvious distinctions must still be made between the two men's approaches. Einstein's maxims of "unity" and "logical simplicity" were contrary to the rather dialectic practice and philosophy followed by Bohr. Furthermore, the same letter in which Einstein identified himself as a "Prinzipienfuchser" also contained his statement that "inductive means will never get you to a sensible theory": just as experiment was increasingly leading the way in studies of the atom by Bohr and Sommerfeld, Einstein allowed his own search for unifying principles to grow more and more removed from induction and experience. Note further that Einstein also typified Max Born, leader of Göttingen's prominent center for quantum theory, as another "virtuoso."

All in all, we may conclude that the practices in the quantum program of the Bohr and Sommerfeld schools, and the philosophical sediments of these practices in methodological pronouncements, stood largely at odds with the direction in which Einstein's post general relativity philosophy of how to do physics developed; it also stood at odds with his actual practice of theory. Such dissonance would have contributed to his distance to the actual study of problems of atomic structure as conducted by the Bohr and Sommerfeld schools, and, one would expect, eventually reinforce a critical position with regards to their further development of the quantum theory. Einstein's growing epistemological separation from experiment, finally, is not only illustrative of his apartness from the atomic quantum program, it also gave him license to remain at a relative distance from efforts in atom building, as well as allowing him to persist in his belief that a unified field theory approach was a reasonable alternative.

Even if Einstein's own work was following different approaches and was largely directed at different problems, he could of course in the early 1920s still be much impressed by the contributions of Bohr and Sommerfeld. In 1922, for example, he complimented Sommerfeld on his gradual "disentanglement" of spectra, and

---

[24] "Es gibt Prinzipienfuchser und Virtuosen. Wir gehören alle drei zu der ersten Sorte und haben (wenigstens wir beide gewiss) wenig virtuosische Begabung. Also Effekt bei Begegnung mit ausgesprochenen Virtuosen (Born oder Debye): Entmutigung. Wirkt übrigens umgekehrt ähnlich." Einstein to Paul Ehrenfest, 18 September 1925, EA 10 111.

praised Bohr's work, in particular his use of the correspondence principle.[25] Yet, regarding the theoretical advances on quanta obtained by then, Einstein would also say that "in my head, at least, things have not become more clear, however large the number of individual facts that one has managed to bring into relationship to one another."[26] The botching together of empirical facts in ad hoc quantum laws was far removed from the esthetic mathematical realms that Einstein gradually came to believe had produced the general theory of relativity, and the kind of unification science that was to produce true understanding; it is thus not surprising that he did not participate in the haphazard elaborations of the Bohr–Sommerfeld atom.

Nevertheless, in the years between the discovery of general relativity and the formulation of matrix mechanics, Einstein did contribute a number of important ideas to the development of quantum theory – so how to assess these? Although Einstein published proposals for "foundational" experiments on light quanta, and signed off on a brief critique of the Stern–Gerlach experiment (co-authored by Paul Ehrenfest[27]), his post-1915 contributions to quantum mechanics were less empirically driven than the theories that came out of the Bohr–Sommerfeld schools. They were not "bottom-up" contributions, following precarious guesswork on the basis of observations. Einstein largely concentrated on unifying theoretical ideas – his work on the Bose–Einstein gas may for instance be qualified as such[28] – rather than exacerbating mysteries. Fundamental innovations could present themselves as they were needed to make the unifying effort succeed: in 1916, for instance, Einstein re-derived Planck's distribution by assuming that radiation was a directional process and that transitions between atomic states were given by probabilistic laws (an assumption that he in fact considered to be a weakness of his setup, expressing a poor understanding of the details of light emission). He could thus bring together Bohr's picture of the atom and Planck's black body law.[29]

It is revealing that none of the work from 1916 on the quantum theory of radiation is mentioned in Einstein's autobiography. According to Abraham Pais, Einstein also considered his work on Bose–Einstein statistics to be of a mere circumstantial relevance – already in 1925 he felt that it did not address the truly fundamental issues.[30] Einstein did not rate these contributions as of the highest value because by 1925 he had already become convinced of a deeper, alternative route for tackling the quantum problem: the field theory route charted by the Solovine schema.

---

[25] See a letter to Sommerfeld of January 1922 (EA 21 402), and Einstein's letter to Ehrenfest of 23 March 1922, EA 10 035.

[26] Einstein to Heinrich Zangger, 18 June 1922, EA 89 500.

[27] See Einstein and Ehrenfest (1922). On light quanta, see Einstein (1921b, 1926a,b).

[28] See Einstein (1924, 1925a,c); see also for example Einstein and Ehrenfest (1923).

[29] See Einstein (1916b).

[30] See Pais (1982), p. 423.

In 1923, in the same year that Einstein and Jakob Grommer's first frustrated attempt at finding a particle solution in the Kaluza theory appeared in print, Einstein wrote a programmatic article, entitled "Does field theory offer possibilities for the solution of the quantum problem?"[31] In Einstein's assessment, the old quantum theory gave a précis of observations, yet true understanding, "a logical foundation,"[32] was still lacking. He believed that the question raised in the title of his article should be answered in the affirmative. By over-determining a set of classical field equations it should be possible to deduce, top-down, the quantum conditions; these, then, would no longer be imposed in an ad hoc fashion, based only on experience. Einstein wrote to Michele Besso that over-determination "is the way in which the non-arbitrariness in initial conditions [i.e. the quantum conditions] should be understood, without abandoning field theory."[33] By following this route, one could hope to arrive "at a truly scientific foundation of the quantum theory."[34] In his letter to Besso, he further admitted to the familiar drawback of his approach that "the connection with experience becomes regrettably ever more indirect," but the possible fruits of his route were too great to leave it unexplored: it was "a logical possibility to do justice to reality without *sacrificium intellectus*."[35]

Physical theory as it had been known until then, including relativity, worked "with differential equations that determine events in a four dimensional spacetime continuum uniquely, if they are known for a spacelike slice," Einstein wrote in his article. "In the unequivocal determination of the temporal continuation of events by partial differential equations lies the method by which we do justice to the law of causality."[36] The practice of field theory and relativity entailed the use of partial differential equations that, in turn, ensured causality. The traditional form of causality was increasingly being questioned by quantum theorists such as Bohr; Einstein however held that a "renunciation of causality *in principle* should only be allowed in the most extreme emergency."[37] Arguing for the retention of causality

---

31 "Bietet die Feldtheorie Möglichkeiten für die Lösung des Quantenproblems?" Einstein (1923b).
32 "[E]ine logische Grundlage," on p. 359 in Einstein (1923b).
33 "So soll die Nichtwillkürlichkeit der Anfangsbedingungen begriffen werden, ohne die Feldtheorie zu verlassen." Einstein to Besso, 5 January 1924, as in Einstein and Besso (1972), p. 197.
34 "[…]zu einer wirklich wissenschaftlichen Fundierung der Quantentheorie." As in Einstein (1923b), p. 364.
35 "[D]er Zusammenhang mit dem Erfahrbaren wird leider immer indirekter. Aber es ist doch eine logische *Möglichkeit*, um ohne *sacrificium intellectus* der Wirklichkeit gerecht zu werden." Einstein to Michele Besso, 5 January 1924, as in Einstein and Besso (1972), p. 197.
36 "Das Wesentliche der bisherigen theoretischen Entwicklung, welche durch die Stichworte Mechanik, Maxwell-Lorentzsche Elektrodynamik, Relativitätstheorie gekennzeichnet ist, liegt darin, dass sie mit Differentialgleichungen arbeitet, welche in einem raumzeitlichen vierdimensionalen Kontinuum das Geschehen eindeutig bestimmen, wenn es für einen raumartigen Schnitt bekannt ist. In der eindeutigen Bestimmung der zeitlichen Fortsetzung des Geschehens durch partielle Differentialgleichungen liegt die Methode, durch welche wir dem Kausalgesetz gerecht werden." As in Einstein (1923b), p. 359.
37 Taken from his criticism of BKS theory: "*Prinzipielles* Aufgeben der Kausalität darf nur in der höchsten Notlage zugestanden werden" ("Bedenken inbez. auf Bohr Cramers," manuscript, 1924, EA 8 076).

and partial differential equations of course also implied arguing for the retention of the conceptual apparatus and familiar mathematics of his own relativity theory, and other field theories. Thus, arguing for causality also meant promoting the field theory program.

Einstein was not alone in imagining a new understanding of quanta through classical field theory. This was to be expected, given the recent successes of general relativity and the frustrations of the old quantum theory; there were other scholars (for example the Dutch mathematical physicist Jan Schouten[38]) who held similar hopes. Even some who were deeply engaged in the quantum program thought highly of Einstein's attempts: Max Born wrote to Einstein in 1925 that the latter's attempts at unified field theories were "much deeper than our petty efforts. I would never dare to trust myself to seek that out."[39] In the same letter, Born announced Heisenberg's latest work on matrix mechanics, his own recent efforts in lattice theory, and further quantum theoretical contributions by his assistants Pascual Jordan and Friedrich Hund.

Fifty years later Werner Heisenberg wrote a historical note on the beginnings of quantum mechanics. He named three centers where the theory had been developed: Copenhagen, Munich and Max Born's Göttingen, where again many quantum theorists, including Heisenberg himself, had spent time for postdoctoral work, and some had done their graduate studies. Revealingly, however, he did not mention Berlin.[40] This suggests that despite Einstein's fundamental contributions to the old quantum theory, quantum theorists no longer regarded him as one of the final theory's founders, nor as a member of one of their collaborative research networks. By and large, the methodology and practices of the quantum program were substantially different from Einstein's changing approaches and beliefs; it can therefore be no surprise that he would promote an alternative, and eventually was looked upon as, indeed, an outsider.

7.2 The formulation of matrix and wave mechanics

Earlier, we have seen the methodological parting of ways between Einstein and the quantum theorists illustrated in the 1920 Bad Nauheim debate, in which Pauli took an operationalist stance to formulate his opposition to the unified field theory program. Einstein did not address Pauli's point directly then and there, but he intimated in a letter to Max Born that he was most reluctant to give up relativity's spacetime continuum: "How could the relative movement of n points be

---

[38] See Goenner (2004), section 1.2.
[39] "[...] tiefer, als unsere Kleinarbeit. Ich würde mir nie zutrauen, darauf auszugehen." Max Born to Einstein, 15 July, 1925, pp. 119–122 in Einstein *et al.* (1969), on p. 121; see also Vizgin (1994), p. 209.
[40] He did so in a manuscript dated 1975; see Heisenberg (1989a), p. 38.

described without the continuum?"[41] Born at first went back and forth between Einstein's and Pauli's positions. In the end, however, he embraced the empiricist approach: as it became clear that the Bohr–Sommerfeld atom – which made use of unobservable electron orbits – would surely fail, Born argued for a heuristic principle that entailed that the real quantum laws must only involve relations between observable quantities such as energies, light frequencies, intensities and phases.[42]

Werner Heisenberg put Born's "observability principle" to use in a first formulation of matrix mechanics: he suggested a way out of the pitfalls of the old theory by reinterpreting kinematical and mechanical relations, and motivated his quantum theoretical reinterpretation by emphasizing its restriction to relations between observable quantities alone.[43] In classical theory the periodic motion of an atom's electron can be written in terms of a Fourier expansion; Heisenberg reinterpreted this sum as an abstract quantum theoretical set that was no longer to be thought of as a description of the electron's orbit. Instead, it was taken to represent an array of numbers – amplitudes and frequencies – that characterized transitions between atomic states. These abstract sets had been suggested by Hendrik Anthony Kramers's formula for the dispersion of light, in which only observable quantities such as radiation frequencies and transition coefficients occurred.[44]

Heisenberg asked how two of such sets multiplied, since, as he pointed out, he could then learn in what way multipole terms in a radiation field arise. Multiplication of the sets had to comply with the empirical Ritz combination principle for the addition of frequencies: $v_{n,n-m'} = v_{n,n-m} + v_{n-m,n-m'}$, where $v_{n,k}$ is the light quantum frequency emitted as the electron decays from the nth stationary state to the kth stationary state. By rearranging the Fourier indices in the product according to this rule, Heisenberg found that in general the quantum theoretical sets should adhere to non-commutative multiplication rules. His proposal for an identification scheme for classical and quantum frequencies would break down if commutative multiplication applied.[45]

It thus seems that Heisenberg did not deduce the puzzling matrix behavior of his abstract sets from some deeply intuited symmetry or axiomatic truth; in his presentation, at least, he rather stumbled on it as he accommodated the Ritz principle. This haphazard way of introducing matrix multiplication – accommodating

[41] "Wie soll man aber die relative Bewegung von n Punkten irgendwie beschreiben ohne Kontinuum?" Einstein to M. Born, 27 January 1920, pp. 42–45 in Einstein *et al.* (1969), on p. 43. Translation as in Mehra and Rechenberg (1982), p. 279.

[42] See Mehra and Rechenberg (1982), pp. 281–283.

[43] See Heisenberg (1925), pp. 261–263 in van der Waerden (1968). For secondary literature, see for example Mehra and Rechenberg (1982).

[44] On the relation between Kramers's result and Heisenberg's work, see Duncan and Janssen (2007), pp. 593–597.

[45] See Heisenberg (1925), pp. 262–266 in van der Waerden (1968), Jammer (1989), pp. 211–213.

the empirical Ritz principle first – could have made Einstein rather uncomfortable. But there were more reasons for discomfort: in matrix mechanics the quantization condition was still imposed from the outset, instead of derived. The theory's construction was further largely based on inductive inferences derived from the correspondence principle – for instance when in Heisenberg's kinematical reinterpretation classical differentials were replaced with discrete differences (as in his formulation of the Thomas–Kuhn sum rule[46]). The strange structure was then a posteriori justified by stating that in principle the theory should only concern itself with observable quantities. This relegated attempts at causal mechanisms – traditional, visualizable explanations for the spectra that for example used partial differential equations – beyond its scope,[47] and the aspiration for a unified, logically consistent theoretical framework beyond its immediate concern. The empirical led the way, as Heisenberg recalled in a 1975 manuscript: "I wanted [...] to trust entirely to the half-empirical rules for the multiplication of amplitude series, which had proved themselves in dispersion theories."[48] The heuristic of mathematical naturalness seems to have been virtually suspended in his efforts: the non-commutative multiplication initially "bothered" Heisenberg, just as matrices were unknown to him[49] – yet this did not undermine his confidence in his preliminary construction.

As matrix mechanics only claimed to concern itself with observable quantities, it appeared to strive to no more than an arrangement of the world of experience. Einstein might have felt that by creed, it remained just above the realm of what he in the Solovine schema referred to as the manifold of experience "*E*," and that the theory made no aspirations to a leap toward a new set of natural axioms "*A*," from which notions such as Bohr's frequency condition or the elementary quantum of charge could be deduced. Heisenberg's construction of matrix mechanics – in which empirically suggested relations took precedence over what might seem more mathematically natural – resembles more the diagram that we drew to capture Einstein's efforts in the Entwurf theory (Figure 2.4), instead of the Solovine schema (Figure 2.2).

So Heisenberg, in the creative phase that produced matrix mechanics, employed a methodological strategy that was the opposite of Einstein's reshaping beliefs. This approach was underpinned by reference to Born's observability principle, and to justify this principle quantum theorists prominently pointed to Mach's philosophy and, particularly, to Einstein's 1905 success in the formulation of special relativity.[50]

---

[46] See Heisenberg (1925), p. 268; see also Jammer (1989), p. 214, Duncan and Janssen (2007), pp. 593–597.
[47] Born and Heisenberg were even of the opinion that with matrix theory spacetime itself was relegated to the macroscopic domain; see Beller (1999), p. 20.
[48] See Heisenberg (1989a), p. 45.
[49] Heisenberg (1989a), p. 45.
[50] See Mehra and Rechenberg (1982), pp. 273–290.

As we know, however, Einstein had long shaken off Mach's point of view and did not have confidence in the particular methodological positions that it had inspired.

Later in life Heisenberg recalled a 1926 visit to Einstein in Berlin. During the visit he had argued that his removal of electron orbits from atomic theory was in line with Einstein's philosophy of special relativity. Einstein, unconvinced by Heisenberg's work, according to the latter replied: "Perhaps I did use such a philosophy earlier, and also wrote it, but it is nonsense all the same."[51] Reference to the "observability principle" only served to remind Einstein of earlier mistaken beliefs – thus further pitching him in a critical position to the emerging quantum theory. Einstein further famously said on this occasion, according to Heisenberg's recollections, that it is in fact the theory which first determines what can be observed: since in the atomic realm the natural laws themselves were still unclear, observation had lost its clear meaning as the interpretations of sensations would necessarily be entirely ambiguous. Heisenberg attributed to these words great influence on his formulation of the uncertainty relations[52] but from the perspective of Einstein's contemporary thought, one could distill a more straightforward message: he may simply have wanted to express that in the atomic realm theory should guide research, and not observation, given the latter's contingencies.

Einstein's response to the work of Heisenberg, and the latter's elaborations, conducted together with Max Born and Pascual Jordan, was at first somewhat ambivalent – Abraham Pais has said that his first reactions reflected a period of vacillation.[53] His letters to Heisenberg on matrix theory are believed to be lost, but Heisenberg's responses to Einstein, particularly a letter that he sent on 18 February 1926, allow some reconstruction of Einstein's initial position. This last letter suggests that Einstein had a number of objections to the theory – yet Heisenberg argued that at least two of Einstein's criticisms might have been due to mistakes in applying the mathematics of matrices. Einstein apparently confused a definition for a commutator $(p_k q_l - q_l p_k)$ with the angular momentum, and may further have made an error when having to apply a non-trivial commutation relation (or perhaps overlooked non-trivial commutation altogether).[54] Einstein, in turn, reported to his friend Michele Besso that he found the mathematics of matrix theory highly

---

[51] Einstein, as cited by Heisenberg in a 1974 lecture, see Heisenberg (1989b), p. 114. On the relation between Heisenberg and Einstein, see also Holton (2005).

[52] See Heisenberg (1989b), p. 114. Einstein also influenced Heisenberg's formulation of the uncertainty relations through the Einstein–Rupp experiments, as they were being discussed by Bohr and Heisenberg; see van Dongen (2007b), pp. 127–129.

[53] Pais (1980), p. 228.

[54] See Werner Heisenberg to Einstein, 18 February 1926, EA 12 172. In another letter, Heisenberg stated that Einstein had found that matrix theory was more suited to the BKS theory's perspective. BKS, however, had been received very critically by Einstein (see his "Bedenken inbez. auf Bohr Cramers," manuscript, 1924, EA 8 076). Heisenberg, in his letter, expressed the hope that energy-momentum conservation and the light quantum would eventually also be accommodated in his quantum mechanics (Heisenberg to Einstein, 30 November 1925, EA 12 171).

unnatural ("a true witches' multiplication table") and believed it to be "sufficiently well protected against proof of falsehood by great complexity."[55] His doubts about the theory would soon prevail; clearly, he would have had higher hopes for his own efforts, centered as they would be on his sense of simplicity and naturalness.

Einstein preferred Erwin Schrödinger's wave approach over matrix mechanics. He initially believed that it was possibly indeed on the way finally to unravel the quantum. In early May 1926 he reported to Besso that the wave theory "smells after deeper truth,"[56] and in a postcard to Schrödinger of 26 April 1926, Einstein applauded the latter's theory, while at the same time expressing his dislike for matrix mechanics.[57]

Einstein's initial attitude to wave mechanics is not surprising for a number of reasons. First, he had stood at the cradle of dual understandings of matter as a result of his contributions to the study of radiation; Einstein had flirted with the idea of some form of a wave-particle duality as early as 1909, then in the context of energy fluctuations in the black body spectrum.[58] In 1926, exactly when he wrote the above postcard to Schrödinger, he was in the middle of his collaboration with Emil Rupp on the wave and particle aspects of light.

Einstein's ideas had been of great importance to Louis de Broglie, who managed to show in 1923 that the Sommerfeld quantum condition can be interpreted as a statement concerning the integer number of wavelengths – of some fictitious wave associated with an internal oscillation of the electron (see also our comments on de Broglie's work in Chapter 5) – that exactly cover the electron's orbit around the nucleus.[59] Einstein repeated de Broglie's conception of assigning a wave to massive particles in a study of the energy fluctuations in a molecular gas, i.e. in one of his papers on the new Bose statistics. De Broglie's ideas inspired Erwin Schrödinger when he picked up the concept of matter waves, and he credited Einstein's reference to de Broglie with convincing him of the importance of the Frenchman's work. Schrödinger eventually rewrote the quantum condition for a particle as an eigenvalue problem for a wave field propagating in configuration space.[60]

A second reason that made it more likely for Einstein to prefer wave mechanics over matrix theory was that Schrödinger's ideals and, to a lesser extent, his approaches in physics were akin to Einstein's. Schrödinger had written against

[55] "Ein wahres Hexeneinmaleins in dem unendliche Determinanten (Matrizen) an die Stelle der kartesischen Koordinaten treten. Höchst geistreich und durch grosse Kompliziertheit gegen den Beweis der Unrichtigkeit hinreichend geschützt." Einstein to Michele Besso, 25 December 1925, on pp. 215–216 in Einstein and Besso (1972).

[56] "Das riecht nach tiefer Wahrheit." Einstein to Besso, 1 May 1926, as in Einstein and Besso (1972), p. 225.

[57] Einstein to Schrödinger, 26 April 1926, EA 22 018.

[58] Einstein (1909a), Einstein (1909b), pp. 482–483; see also for example Klein (1964), pp. 5–15.

[59] See Jammer (1989), pp. 247–249.

[60] For Einstein on de Broglie, see Einstein (1925a), p. 9; on the influence of de Broglie on Schrödinger, see Schrödinger to Einstein, 23 April 1926, EA 22 014; on Schrödinger, see for example Wessels (1979); Kragh (1982) and Joas and Lehner (2009).

a Machian methodology, admitted to having been disinclined to become deeply involved in atomic theory because of its ad hoc nature and shared Einstein's dislike for matrix theory.[61] Earlier, in 1922, he had tried to connect the quantum conditions to Weyl's unified field theory,[62] and his new wave theory approach promised again a more direct connection between quantum theory and the unified field theory program.

Wave mechanics seemed to entail a certain visualizability, or intelligibility, of atomic processes as wave phenomena. For Schrödinger, this *Anschaulichkeit* had in fact been an important motivation for constructing his theory.[63] In a later piece, "Physics and reality," when comparing the Heisenberg and Schrödinger approaches, Einstein said that the latter "lies closer to the physicist's method of thinking."[64] Einstein reputedly described his own thinking as an "associative play" of "more or less clear images" of a "visual and some of muscular type"[65] – from that perspective it is to be expected that he would prefer work inspired by a striving for *Anschaulichkeit* over the opaque matrix theory. Just as Einstein's preference was to be expected, Schrödinger could also count on a critical reception from the majority of the quantum theorists: the emphasis on *Anschaulichkeit* soon brought him in conflict with Bohr and the adherents of the observability principle from Göttingen.[66]

The wave theory looked like it might leave a strict and visualizable spacetime causality intact; in line with Einstein's programmatic paper of 1923, it maintained differential equations and, as said, held out the promise of a more direct link with field theories and relativity – as we pointed out in an earlier chapter, Schrödinger had actually earlier considered a relativistic field theory that however failed to give the right hydrogen spectrum.[67] *Anschaulichkeit* and causality of course also resonated with Einstein's desires for simplicity and naturalness. The theory even suggested the possibility of a particle-like solution that could represent electrons or perhaps other particles; Schrödinger made recommendations for solving this problem, but had to admit that superpositions of solutions to his wave equation would produce unstable wave packets instead of stable particles. Finally, he thought that the wave equation removed some of the ad hoc character of the quantum conditions, since his equation was a quantum equation that could be motivated by a variational procedure.[68]

Already quite early on Einstein also had reservations regarding wave mechanics, despite all its positive attributes. He found it particularly objectionable that

---

[61] Wessels (1979), pp. 311–313; for Schrödinger on matrix theory, see Kragh (1982), p. 178.
[62] See Wessels (1979), pp. 321–322. This work likely made Schrödinger more receptive to de Broglie's ideas.
[63] On *Anschaulichkeit* in Schrödinger's work, see de Regt (1997).
[64] As in Einstein (1936a), on p. 345 in Einstein (1994).
[65] As in Hadamard (1949), pp. 142–143.
[66] For that conflict, see de Regt (2001), pp. 252–255.
[67] On this issue, see Kragh (1982).
[68] On his attempt to form particle solutions, see Wessels (1979), pp. 326–327, and on the removal of the ad hoc character of the quantum conditions, see pp. 331–332 and Kragh (1982), p. 164.

the waves propagated in configuration space instead of real space. As he wrote to Arnold Sommerfeld: "If only the undulatory fields introduced there could be moved from the n-dimensional coordinate space to the 3 or 4-dimensional space!"[69] To justify his theory, Schrödinger had used the analogy between geometrical optics and the Hamilton–Jacobi theory of particle mechanics in his publications[70] – this analogy gave the theory its formal appeal, but on its basis it could be expected that the Schrödinger wave function would propagate through configuration space. As Einstein's letter to Sommerfeld suggests, however, the configuration space description stood in the way of a straightforward realist interpretation of the wave-fields: if these had been propagating through four-dimensional spacetime, the connection with his own fields and attempts to find particle solutions in field theories would have been easier to establish. In Schrödinger's theory, however, the number of dimensions grew with the number of particles.[71] This aspect would of course decrease the theory's "mathematical naturalness." Furthermore, a few months before Einstein's letter to Sommerfeld, Born had already employed wave mechanics to formulate a first quantum-mechanical theory of particle scattering, and he proposed to interpret the square of the particle's wave amplitude as its probability to assume a given position; such an interpretation would further undermine the possibility of connecting the Schrödinger wave to the fields of Einstein's unification attempts.[72]

Soon Einstein would take up his position of lasting dissent. Even before many of the ingredients of the Copenhagen interpretation had been established, he expressed that in his view the new quantum mechanics did not provide a proper explanation for such things as for example the electron spin – there was no hope of understanding this concept's "inevitability *from within*," and in more general terms, Einstein held that the theory did not smell "after reality."[73] In late 1926, after it had been argued that wave and matrix mechanics were in fact mathematically and empirically equivalent,[74] he wrote to Max Born:

Quantum mechanics is certainly imposing. But an inner voice tells me it is not yet the real thing. The theory says a lot, but does not really bring us closer to the secret of the Old One.[75]

[69] "Wenn die dort eingeführten undulatorischen Felder nur aus dem n-dimensionalen Koordinatenraum in den 3 bezw. 4 dimensionalen verpflanzt werden könnten!" Einstein to Sommerfeld, 21 August 1926, as in Einstein and Sommerfeld (1968), p. 108.

[70] On the role of the optical-mechanical analogy in Schrödinger's creative process, see Joas and Lehner (2009).

[71] For a related viewpoint, see Einstein (1933a), on p. 302 in Einstein (1994).

[72] On the relation between Einstein's thinking about wave-particle duality and Born's interpretation, with a particular emphasis on the Einstein–Rupp experiments, see van Dongen (2007b), pp. 126–127.

[73] See Einstein to Arnold Sommerfeld, 21 August 1926: "[...] riechen mir aber nicht nach der Wirklichkeit"; "[...] einstweilen ist wenig Hoffnung, seine Notwendigkeit *von innen heraus* zu begreifen," p. 108 in Einstein and Sommerfeld (1968), emphasis in original.

[74] Philosophers of science today however debate whether Schrödinger's equivalence proof of March 1926 was in fact insufficient; see Muller (1997), Perovic (2008).

[75] "Die Quantenmechanik ist sehr achtung-gebietend. Aber eine innere Stimme sagt mir, dass das doch nicht der wahre Jakob ist. Die Theorie liefert viel, aber dem Geheimnis des Alten bringt sie uns kaum näher." Einstein to Max Born, 4 December 1926, on p. 129 in Einstein *et al.* (1969).

Despite all the advances made, true understanding had not been achieved; in fact, in that respect the theory was no improvement over the old quantum theory – the empiricist, "bottom-up" methods of Born and his fellow quantum theorists had yielded no more than a phenomenological theory.

Einstein continued his letter by famously saying that God "does not play dice." He further suggested again that waves in $3n$ dimensions were not quite natural – yet these elements of his letter have a certain expedient quality, even if they reflect the realist and deterministic stance that Einstein later became strongly identified with. The most important, yet still understated, motivation for Einstein's disengagement was largely elsewhere: he wished to persist in his field theory approaches. Immediately after stating that he did not believe that God played dice, Einstein informed Born that he was trying to derive the geodesic equations for singular particles from the general relativistic field equations – one of his subplots in the larger plan of the unified field theory approach to the quantum. Similarly, he wrote to Hendrik Antoon Lorentz in early 1927:

The quantum theory is completely Schrödingerized and has a lot of practical success because of it. But this cannot be the description of what really happens. It is a mystery. It turns out that the unification of gravity and Maxwell theory is achieved in a completely satisfactory manner by the five dimensional theory (Kaluza–Klein–Fock). I am curious about what you have to say about that.[76]

The, rather sketchy, disqualification of quantum mechanics was contrasted with the promise of the Kaluza–Klein theory.

Einstein's actual practice and outlook in physics was not much affected by the developments of 1925 through 1927, developments that to others, like Niels Bohr and Werner Heisenberg, were truly revolutionary.[77] He saw continuity rather than radical change in the quantum program: new and old versions of the theory were both at heart only empirical, phenomenological constructs and were equally insufficient. As we have seen, Einstein was not himself involved in the atomic studies

---

[76] "Die Quanten-Theorie ist ganz verschröderingert und hat viel praktischen Erfolg davon. Aber dies kann doch nicht die Beschreibung des wirklichen Vorganges sein. Es ist ein Mysterium. Es zeigt sich, dass die Vereinigung von Gravitation und Maxwell'scher Theorie durch die 5 dimensionale Theorie (Kaluza–Klein–Fock) vollständig befriedigend geleistet wird. Ich bin neugierig, was Sie dazu sagen werden." Einstein to Hendrik Antoon Lorentz, 16 February 1927, p. 649 in Kox (2008).

[77] Suman Seth (2007) has similarly argued that Arnold Sommerfeld did not experience a sense of crisis before 1925, and did not see a revolution take place in 1925–1927; in the case of Sommerfeld this was due to the fact that he was focused on particular problems, rather than contemplating principles. Seth further offers thoughts on whether there really was a Kuhnian sense of crisis and subsequent revolution in physics around 1925. He suggests this may have been true for some historical actors, but not for others, convincingly arguing the latter case for Sommerfeld's group in Munich. In our assessment, Einstein, on the issue of revolution, would be just such an exception. Einstein was further an exception to the controversial "Forman thesis." This claims that the physicists' sense of crisis and a-causal bent resonated with a broader Weimar malaise and culture (Forman 1971) – Einstein however saw great hope in the creation of the Weimar republic (see for example his letter to Paul Ehrenfest of 12 September 1919, Doc. 103, pp. 154–155 in Buchwald *et al.* (2004)), and did not relinquish on causality.

that led to the first formulation of the new theory, and decreasingly identified with how atomic physicists built their theories. In particular, his earlier experiences with the general theory of relativity made him increasingly insistent on the feasibility of a different approach. He remained convinced that his program – his top-down approach, based on maxims of simplicity and naturalness, and familiar concepts and tools such as relativistic fields and their partial differential equations – was a promising alternative that in the end would carry the day. It would yet give a true theory of reality – beyond the mere phenomenological, and observant of a more intuitive notion of causality. That theory was to be found by following his route, his philosophy, his method – a tried and tested method – as he would often express in the years to come.

7.3 Einstein and quantum mechanics

In what follows, we will briefly address Einstein's attitude regarding the fully developed quantum theory. Einstein felt that the developed theory of quantum mechanics did not fit the epistemology of the Solovine schema, just as the methods used to formulate the theory would not have corresponded to the recipes of the Spencer lecture; these circumstances emphasized that the theory had not yet given a proper explanation of the quantum, and strengthened his resolve that his methodological prescriptions would provide a viable alternative. There were numerous elements of the theory that further convinced Einstein of these contentions – most importantly, the problems that he perceived the probabilistic description produced for such notions as causality, determinism and the completeness of physical theory; these problems would also likely be absent in his desired field theory account of the quantum.

Einstein developed his criticisms in an exact and explicit manner on a number of occasions, for example in his famous 1935 "EPR" argument, initially published together with Boris Podolsky and Nathan Rosen, and later presented anew in other publications.[78] His criticism has of course been perused with great care by many historians and philosophers and is, we presume, familiar to our readers.[79] Our historical and methodological perspective should not be taken to supersede the familiar elements of Einstein's critique of quantum mechanics; as the above has already suggested, it is rather a natural complement to those elements. Our focus in this section will be to illustrate that Einstein's familiar criticism fitted in well with the content and aspirations of his own contemporary research program.

[78] See Einstein *et al.* (1935), Einstein (1936a). For a discussion of the various manifestations of the EPR argument in Einstein's work, see Fine (1996), Chapters 3, 4 and 5; for a discussion that centers on the issue of "separability," see Howard (1985).

[79] A recent extensive overview is contained in part B of Home and Whitaker (2007).

Many of the reservations and objections that Einstein expressed regarding quantum theory intended to express that it still only provided a preliminary and phenomenological, bottom-up description; in 1940, he found that the theory owed "its origin" to "facts of experience," but however compelling these may be, physicists still did not have a "general theoretical basis" that can function as a "logical foundation."[80]

Initially, for a few years, Einstein argued for the inconsistency of the theory. In between sessions at the 1927 and 1930 Solvay conferences, he proposed thought experiments that were to exhibit incongruities – yet after they were repeatedly rebuffed by Niels Bohr, Einstein settled on admitting to the theory's consistency.[81] However, it still neglected the gravitational force and the general theory of relativity, and this, to Einstein, again emphasized its preliminary status.[82] Such a judgment was of course to be expected in light of the fundamental relation in his thinking between unification and explanation.

On the few occasions that Einstein dared to imagine what a quantum gravity theory might possibly look like, he ventured that it would perhaps dispense of the spacetime continuum and its accompanying differential equations, but he found that "instinct rebels against such a theory."[83] The well-nigh impossibility of formulating a quantum gravity theory strengthened his conviction that quantum mechanics had to be off track. Presumably, such a theory would somehow have to involve the quantization procedures of quantum field theory, but, as we know, Einstein felt these to be at odds with his maxims of mathematical naturalness and simplicity, as quantization would only give an indirect description: "I see in this method only an attempt to describe relationships of an essential nonlinear character by linear

---

[80] See Einstein (1940), pp. 491–492.

[81] On these discussions, see Pais (1982), pp. 444–448; a first hand account by Bohr of the Solvay discussions is Bohr (1949). The above version of events has been termed the "standard history" by Don Howard, and he believes that an alternative history is more likely; in this version Einstein would have maintained as his principal objection to the quantum theory its failure to satisfy a "separability" principle from as early as 1925. The EPR argument would further have been intended to illustrate exactly this failure. On this account, Einstein would not have altered his line of argument as indicated above; see Howard (1990b), in particular p. 62. On the 1930 debate with Bohr, see also Dieks and Lam (2008).

[82] See for example Einstein (1936a), on p. 351 in Einstein (1994); Einstein wrote to Schrödinger on 17 June 1935 that the statistical nature of quantum theory prohibited its inclusion of general relativity (EA 71 836). He further wrote to Max von Laue (letter of September 1950) that he thought it hopeless to generalize a special relativistic quantum theory to a general relativistic theory, and informed a certain K. Roberts (letter of 6 September 1954) that he therefore had not really studied quantum field theory: see p. 378 in Stachel (1986). The sentiment that fundamental theories that ignored gravity could only have a preliminary status is also in evidence in Einstein (1950), p. 16.

[83] Einstein to Paul Langevin, 3 October 1935, as in Stachel (1993), on p. 150 in Stachel (2002). For Einstein's thoughts on discretized gravity theories, see further for example Einstein to H. S. Joachim, 24 August 1954, cited in Stachel (1986), on p. 396 in Stachel (2002), and Stachel (1991), on p. 416 in Stachel (2002); see also Stachel (1993) for more examples. For Einstein's disbelief in the feasibility of a dismissal of spacetime, see further his letter to Schrödinger of 17 June 1935: "I consider the renunciation of a spatio-temporal setting for real events to be idealistic-spiritualistic" ("[ich finde] den Verzicht auf eine raum-zeitliche Erfassbarkeit des Realen idealistisch-spiritistisch"), on p. 68 in Fine (1996); EA 71 836. See also Einstein (1936a), on p. 351 in Einstein (1994).

methods."[84] If reconciliation was at all attainable, he expected the quantum to yield to gravity, not the reverse: "all attempts to obtain a deeper knowledge of the foundations of physics seem doomed to me unless the basic concepts are in accordance with general relativity from the beginning."[85]

Quantum theorists of course formed their own intuitions about the mathematical simplicity and naturalness of their theory. Particularly important in this regard were contributions by Paul Dirac and John von Neumann, who produced formalizations of the theory, captured in widely disseminated books that were published in 1930 and 1932 respectively. Einstein knew both volumes,[86] yet never seems to have internalized their perspectives. As we saw, he felt the quantum theory's use of configuration space to be unnatural, and a similar objection would easily apply to the Hilbert space formalism: Einstein found it problematic that the theory made "no claim to be a mathematical model of the atomic structure."[87]

He had not been involved in the development of the theory's formalism and had in the previous decade trained himself thoroughly in a different type of mathematics, namely the differential geometry of relativity. It should thus not be a surprise that formalizations and re-conceptualizations of the quantum theory would not convince him, even if quantum theorists would eventually have recognized a Solovine schema or a sense of mathematical naturalness in their theory. To expect the same of Einstein would be to overlook his historicity: his practice in theory had been increasingly engrained with the mathematics of classical fields, which found its motivation and justification in the recollection of 1915. In this way Einstein had effectively come to hold classical fields and partial differential equations to be the only proper categories with which to seek an account of the puzzling quantum. Unsurprisingly, his well known objections to quantum theory that were centered on the issues of completeness, determinism and causality fitted well that implicit preference.

Retaining the familiar field concepts from general relativity thus seemed in Einstein's physics in effect to be implied if one were to follow a "mathematically natural" route. This meant that a justification for his negative attitude towards quantum theory could be given by pointing to the methodological recipe: emphasizing mathematical naturalness thus also implied dismissing quantum theory. As

---

[84] As in Einstein (1955), p. 165.

[85] As in Einstein (1950), p. 16. Don Howard has taken Einstein's various versions of the EPR argument primarily to point out a problematic non-separability of two systems that interacted in the past (i.e. though spatially separated, the systems do not possess separate real states; see also note 81). Einstein would not allow such non-separability, Howard (1985), pp. 196–197, argues, and neither would field theories such as general relativity. Howard's interpretation is of course in accordance with the cited comment, which upon that reading would further show a close coherence between Einstein's insistence on field theory and his criticism of the quantum theory. The above remark should however not be construed as support of Howard's perspective, as its actual context is different: Einstein pointed instead to his experience with formulating general relativity to motivate his contention.

[86] According to his secretary, Helen Dukas, Einstein always kept Dirac's book (1930a) close at hand; see p. 261 in Jost (1980). For Einstein's knowledge of von Neumann's work (1932), see for example Valentin Bargmann to Einstein, 26 July 1940, EA 6-209.

[87] As in Einstein (1933a), on p. 302 in Einstein (1994).

we have repeatedly seen, however, his methodological beliefs in turn also motivated and directed his alternative attempts at a description of the quantum, and thereby his retention of classical fields as well; this pursuit led to a hardened resistance to succumb to the increasing successes of quantum theory.

Furthermore, despite what anyone steeped in the formalizations of Dirac or von Neumann may have found, he had seen quantum theory grow out of experience while he elaborated unified fields: even when confronted with a fully developed theory, Einstein would not forget its bottom-up empiricist origins, and its creation by methods different from his – he would thus not easily have recognized the Solovine schema in the fully developed theory. Einstein had already become committed to the field theory program before quantum theory had matured, and the objections he raised, like the EPR argument, should thus also be understood as intended to advertize his alternative program.

In the case of Einstein's discomfort with phase space and Hilbert space descriptions, as in other cases, his desire for simplicity and mathematical naturalness coincided with difficulties in giving the theory a straightforward realist interpretation. The quantum theory made "no attempt to give a mathematical representation of what is actually present or goes on in space and time," and in this respect, it differed "fundamentally from all previous theories of physics."[88] The desire for simplicity and naturalness also coincided with his rejection of probabilities. In Oxford, immediately after complaining about the theory's specific usage of abstract configuration spaces, Einstein indeed continued with: "I cannot but confess that I attach only a transitory importance to this [i.e. probabilistic] interpretation. I still believe in the possibility of a model of reality – that is to say, of a theory which represents things themselves and not merely the probability of their occurrence."[89]

As said, Einstein's dislike for quantization was a clear expression of his methodological convictions: quantization implied that the quantum rules and probabilities were not derived from some deeper underlying set of natural assumptions, but instead, they were taken on from the outset, which would result in a phenomenological, bottom-up theory. In a letter from 1932 on the projective Kaluza–Klein and teleparallel unified field theories, he remarked to Wolfgang Pauli (with a nod to Cato the Elder):

By the way, I do not say
   "probabilitam esse delendam," [Probability needs to be removed]
but rather
   "probabilitam esse deducendam," [Probability needs to be deduced]
which after all is not the same.[90]

88 Einstein (1940), p. 491.
89 As in Einstein (1933a), p. 302 in Einstein (1994).
90 "Übrigens sage ich nicht "probabilitam [sic] esse delendam," sondern "probabilitam [sic] esse deducendam," was doch nicht dasselbe ist." Einstein to Pauli, 22 January 1932, EA 19-170, pp. 109–110 in von Meyenn (1985).

In this instance the methodological discomfort took precedence over Einstein's critique of probabilities – indeed, in light of these comments, it would be too simple minded to qualify Einstein's position as one that simply held that "God does not play at dice."[91]

Clearly, one needs to engage Einstein's positive program in field theory in order to get a full view of his position regarding quantum mechanics. The reverse is of course also true: the field theory approach often appeared in unison with Einstein's stance on quantum theory. As we have seen, he expected much from singularity-free solutions of generalized field equations as possible particle models. In his programmatic article on "Physics and reality," for example, his critique of quantum theory was immediately followed by a statement of the promise of particle solutions: a positive attribute of such solutions would be that the behavior in time of these particles would be "determined solely by the differential equations of the field"[92] – and thus entail a less problematic notion of causality, Einstein likely thought.

Pauli, in a letter in 1948, expressed disbelief that such non-singular particle solutions would allow descriptions of experience that could somehow avoid probabilities at a more fundamental level.[93] Einstein remained unconvinced but earlier, in 1933, he had been willing to acknowledge as a lesson to be learnt from Heisenberg's uncertainty principle that "complete localization of particles in a mathematical model" may have to be given up. Such non-localized models were "perfectly thinkable. For instance," Einstein continued, "to account for the atomic character of electricity, the field equations need only lead to the following conclusions: A region of three dimensional space at whose boundary electrical density vanishes everywhere always contains a total electrical charge whose size is represented by a whole number."[94] We have seen that Einstein tried to construct these kinds of soliton solutions in many of the theories he studied – in particular in the semivector theory that was under scrutiny when he spoke the above words in Oxford in 1933.

Einstein's letter to Pauli cited earlier implied that he felt that probabilities are not themselves fundamental ("Probabilitam esse deducendam"), in contrast to the Copenhagen interpretation. This is reminiscent of a so-called ensemble interpretation of quantum theory. In Einstein's words: "The ψ function does not in any way describe a state which could be that of a single system; it relates rather to many systems, to an 'ensemble of systems' in the sense of statistical mechanics."[95]

---

[91] See for this qualification for example Heisenberg (1989b), pp. 116–117.

[92] As in Einstein (1936a), on p. 352 in Einstein (1994).

[93] Pauli to Einstein, 21 April 1948, EA 19 187.

[94] See Einstein (1933a), on p. 302 in Einstein (1994).

[95] As in Einstein (1936a), p. 349 in Einstein (1994). Einstein often referred to this interpretation as the "Born interpretation"; he told Gregory Breit however that he was not certain whether Born had consistently represented this interpretation; see Einstein to Breit, 2 August 1935, EA 6-173.

Einstein likely believed that the actual state of the individual system is such that its observables have a definite value, and measurement reduces the original quantum mechanical wave-function to that of a subensemble with appropriate values for the observables. He thought of quantum mechanics as analogous to a statistical mechanical theory, "only with the difference that here we have not found the equations corresponding to those of classical mechanics."[96]

Einstein's adoption of an ensemble interpretation reflected his conviction that the theory had to be considered an incomplete description of reality, as also followed from the EPR argument. He admitted the correctness of the theory's predictions and its logical consistency, but from the position of incompleteness denied its ultimate authority in representing the "real external world."[97] In a 1939 letter to Erwin Schrödinger, Einstein said that he found the latter's cat experiment the best example to illustrate incompleteness; current physics simply lacked theories that could give complete descriptions of reality (even if the situations described, even when finally captured by a complete theory, might not be "observable in their entirety for single cases"). He ended his letter by pointing out that he had arrived at a "truly mathematical theory"[98] – this would have been the Einstein–Bergmann–Bargmann version of Kaluza–Klein theory. Even if Einstein further admitted to Schrödinger that this theory was hard to test experimentally, the larger message was clear: incompleteness was to be overcome through his methodologically sound endeavors in unified field theory.

Einstein was of course elaborating field theories with the same vision on reality and physical theory as the vision that inspired his position on quantum mechanics; the ways in which he attempted to retrieve quantum theory from classical field theory might therefore limit the way in which he wanted to interpret quantum mechanics. His attempts in field theory render an ensemble interpretation rather natural. The observables that Einstein constructed in unified field theory were usually the familiar observables, straightforwardly defined through classical field theory.[99] This would imply that the actual state of an individual particle has to be dispersion free, as seems natural in Einstein's interpretation as outlined above.

There are however indications that Einstein may not have thought of an ensemble interpretation as the ultimately satisfactory way to think about the relation between quantum probabilities and the actual world.[100] It has been pointed out

---

[96] Einstein to Paul Langevin, October 3, 1953 as in Stachel (1986) on p. 395 Stachel (2002). On the issue of whether Einstein held that his ensemble interpretation was, in fact, Gibbsian (rather than "minimal"), see Home and Whitaker (2007), pp. 189–196.

[97] "Real external world" as in Einstein (1936a), on p. 319 in Einstein (1994).

[98] Einstein to Erwin Schrödinger, 9 August 1939, EA 22 061.

[99] See for instance his electric current in equation (6.17); our expression for charge in (6.13) might serve as another illustrative example.

[100] See Fine (1996), pp. 40–59, Home and Whitaker (2007), pp. 194–196.

that he might have wanted to arrive eventually at observables different than the observables familiar from classical theory; the relation between actual reality and probabilities would then be more complex than captured by the quantum formalism in an ensemble interpretation.[101] Furthermore, in 1935 Schrödinger presented arguments against ensemble interpretations such as the interpretation presented above.[102] Einstein was aware of Schrödinger's criticisms, and he never indicated how his interpretation might avoid them; this suggests that he would have thought of an ensemble interpretation as only a preliminary and expedient perspective. In fact, he never really gave a detailed account of how exactly his particular interpretation would relate the quantum formalism to the underlying reality. This inarticulateness is reflected in his qualification that quantum mechanics would, with respect to a future complete theory, take at most an "approximately analogous position"[103] to that of statistical mechanics within the framework of classical mechanics. It may be that Einstein only presented the elements of an ensemble interpretation to draw attention to the issue of incompleteness – this, in turn, of course being another way of whetting the appetite for possibly complete theories, that is, his unified field theory attempts.[104]

As we have seen, Einstein was willing to draw as a lesson from the Heisenberg uncertainty relations that localization in models for individual particles may have to be given up. In an ensemble interpretation, however, the uncertainty relations do not apply to the actual state of an individual particle. It seems likely that Einstein may have wanted to reinterpret the uncertainty relations when he found his desired soliton solution. That solution proved elusive, however, and this stands in the way of a reconstruction of how he might have intended to re-assess the Heisenberg relations. The classical field theory definitions for observables actually employed by him do nonetheless suggest that he expected the soliton particle state to be dispersion free, and John Bell has argued that Einstein's commitment to field theories should indeed be seen as an implicit commitment to the hidden variable program.[105]

---

[101] See for this point Fine (1996), p. 58, Whitaker (1996), p. 242. It should be pointed out however that the documentary evidence to support this contention is, in our opinion, rather scant; only a single source is quoted by Fine that directly supports the idea that this was an element of Einstein's thinking: Einstein to Schrödinger, 31 May 1928, cited on p. 18 in Fine (1996). The date of this source is still relatively early in both Einstein's reaction to the quantum theory and his field theory program. The sustained use of the familiar observables defined through classical field theory, as discussed above, suggests that a desire to dismiss classical concepts need not have been of great priority to Einstein.

[102] For example, he argued that if only a certain number of quantum mechanically allowed values of angular momentum occur for a particular choice of coordinate axis, in an ensemble interpretation one cannot change the axis, even though its choice should really be arbitrary: this would lead to values of angular momentum that are not allowed. See Home and Whitaker (2007), p. 193.

[103] As in Einstein (1949b), p. 672; see also p. 196 in Home and Whitaker (2007).

[104] See also Fine (1996), p. 58, Home and Whitaker (2007), pp. 193–196.

[105] See Bell (1987), pp. 89–91.

Be that as it may, Einstein did not receive warmly two explicit examples of a hidden variable theory, namely the theories of Louis de Broglie and David Bohm. These theories constructed hidden variables for the Schrödinger wave-function – they were theories that aspired to complete quantum mechanics "from within," to paraphrase philosopher Arthur Fine.[106] Einstein did not support de Broglie at the initial presentation of his model at the 1927 Solvay conference and also received Bohm's work critically. In 1954 Bohm communicated his latest ideas to Einstein, yet they could not convince the latter: he wrote to Bohm that in his opinion, "we are still quite remote from a satisfactory solution of the problem."[107]

Bohm, in his reply, gave his motivation for studying hidden variable theories: he believed that there was a subquantum level that again would be deterministic and continuous. He also explained why he did not have high expectations of Einstein's approaches. Bohm felt that one should distinguish between macroscopic and microscopic laws, and he suggested that Einstein was trying to find the latter through the former, which he deemed not very promising.[108] Einstein answered: "I do not believe in micro and macro laws, but only in structure laws that lay claim to a general and strict validity."[109] Already in 1921, Pauli had criticized the unified field theory program in his encyclopedia article by pointing out that the strength of the gravitational versus the electromagnetic force differed by some twenty orders of magnitude.[110] This suggested that there was not much to be learned about the interactions in the atom from unifying gravity and electromagnetism. The above remark illustrates why Einstein could dismiss such a position: he did not think in terms of laws that applied at different scales. Instead he believed that there should be just one theory that should give predictions for any scale. Mathematical structure overrode empirical contingency.

Einstein had more to say to Bohm about such structure laws:

I believe that these [structure] laws are *logically simple* and that the faith in this logical simplicity is our best guide, in the sense that it suffices to start from relatively little empirical knowledge. If nature is not arranged in a way corresponding to this belief, then there is no hope at all to arrive at deeper understanding. [...]

This is not an attempt to convince you of anything. I just wanted to show you how I came to my position. The realization that in a half empirical way one could never have arrived

---

[106] As in Fine (1996), p. 57; see also Whitaker (1996), p. 239.
[107] Einstein to Bohm, 28 October 1954, EA 08 050. In 1927 Einstein had nevertheless developed his own hidden variable theory but retracted its publication, as he believed his scheme was flawed; see Belousek (1996). Belousek (1996), p. 453, holds that this experience contributed to Einstein's negative attitude to de Broglie–Bohm type theories. On Einstein's reception of de Broglie's hidden variable ideas, see Cushing (1994), pp. 118–120. On the wider reception history of the Bohm theory, see Freire (2005).
[108] Bohm to Einstein, 14 November 1954, EA 08-052.
[109] "Ich glaube nicht an Mikro- und Makro-Gesetze, sondern nur an Struktur-Gesetze die allgemeine strenge Gültigkeit beanspruchen." Einstein to Bohm, 24 November 1954, EA 08-054; also in Stachel (1991), on p. 409 in Stachel (2002).
[110] See Pauli (1921), on p. 205 in Pauli (1958).

at the gravitational equations for empty space had a particularly strong influence on me. In that case, only the viewpoint of logical simplicity can be of help.[111]

This brings us back, full circle, to Einstein's methodology and his struggles with the creation of general relativity. Again we see him relate that this was the defining moment for his epistemology; he had arrived fully at the position that it was mathematics that had delivered him the theory, and he squarely engaged that position in his debate on quantum theory. It foremost justified the contention that an alternative route can and ought to be explored – it appears that Bohm's hidden variables did not concern Einstein much from the perspective of his alternative route. In fact, from that perspective all quantum mechanical theory looked like a premature aberration.

7.4 Conclusion: the quantum and the practice of field theory

"Our problem is that of finding the field equations for the total field," Einstein observed in his autobiographical notes – he further said that he did not expect quantum theory to be a good starting point for the "theoretical foundation of the physics of the future," and argued that the EPR argument had shown that the "statistical character of the present theory" was a "necessary consequence of the incompleteness of the description of the systems in quantum mechanics."[112] As these quotes exemplify, Einstein's negative position regarding quantum theory was wedded to his positive program in field theory. They were two sides of the same coin, as even his choice of words suggests: "incompleteness" was to be overcome by finding field equations for the "total" field. As early as 1926 Einstein objected to the new quantum theory, and hinted at the continued promise of the field theory program. He had already been strongly committed to that program for a number of years, while his objections to quantum theory still had to crystallize in his thought. In the decades to come, his criticism became more pointed, just as his defense and advertisement of the field theory approach became more pronounced; both positions developed in tandem, as Einstein's research immersed itself ever deeper in unified field theories.

As we have tried to argue, and as Einstein often emphasized, there is a particular dimension in his personal history that helps to explain his attitude: the discovery of general relativity and the epistemological reorientation that Einstein underwent

---

[111] "Und ich glaube daran, dass diese [i.e. Struktur-] Gesetze *logisch einfach* sind, und dass das Vertrauen an diese logische Einfachheit unser bester Wegweiser ist, derart, dass es genügt von verhältnismässig wenigen empirischen Erkenntnissen auszugehen. Wenn die Natur nicht diesem Glauben entsprechend eingerichtet ist, dann haben wir überhaupt keine Hoffnung, tieferes Verstehen zu erlangen. [...] Dies ist kein Versuch, Sie von irgendetwas zu überzeugen. Ich möchte Ihnen nur zeigen, wie ich zu meiner Attitude gekommen bin. Besonders stark hat auf mich die Erkenntnis gewirkt, dass man auf einem halb-empirischen Wege niemals zu den Gravitationsgleichungen des leeren Raumes hätte gelangen können. Nur der Gesichtspunkt der logischen Einfachheit kann da helfen." Einstein to Bohm, 24 November 1954, EA 08-054.

[112] As in Einstein (1949a), pp. 81–89.

following that experience. Theories ought to be found by following the method outlined by the Spencer lecture and Solovine schema; quantum mechanics was not developed according to these lines of inquiry, so the appropriate route for the construction of a theory for the quantum had not yet been properly explored. As we have seen, however, the recollection of 1915 and its methodological lesson would in turn also be informed by the field theory program, and employed prominently to justify and promote it; Einstein evidently used his objections to the quantum theory in a similar way.

Even if some quantum theorists might later claim that the theory had introduced its own notions of mathematical naturalness or logical simplicity, early on Einstein had already decoupled from the quantum program to such a degree that it can be no surprise that these notions were not to resonate with his. He had lacked close involvement with that strand of the old quantum program that developed the full theory from 1925 through 1927, as his themes had been different from the intricate studies of the atom that were directed from Copenhagen, Göttingen and Munich. The difference between Einstein and the quantum theorists was further reflected by the contrasting roles that experience and experiment played in their respective ways of doing theoretical physics. It was finally also expressed by the distinction between the particular kinds of mathematics employed by Einstein and the quantum theorists; in the end, they would simply not share the same sense for what would be a physical theory's naturalness or simplicity.

The differing roles of experiment were not just reflected in the practice of theory construction, but also ideologically: when quantum theorists expressed their allegiance to Mach, Einstein could only disagree, and he was not willing to surrender familiar notions of causality for the expediency of agreement with experiment. As quantum mechanics gradually took its shape, Einstein had already set on a different track: the track that the experience of general relativity had put him on, the track that aspired to a unified and complete representation of reality through a mathematically natural field theory. As he had told Max Born as early as 1926, "an inner voice" told him that quantum mechanics "is not yet the real thing." His track, charted by the unified field theory methodology and the changing recollection of 1915, was to lead beyond the merely phenomenological of the quantum theory, beyond the "*S*" and the "*E*" from the Solovine schema, to a natural, simple and unified conception of reality.

Conclusion

The aim of this book has been to understand Einstein's development and see the historical coherence in his later attitude to physics. The key that unlocks the later Einstein is the road by which he arrived at the field equations of general relativity, "the most joyous moment of my life." With superficial hindsight, mathematical intuition and deduction had played the essential creative role, whereas his "physics first" arguments appeared to have been a hindrance. As we saw, his epistemology and methodology gradually changed accordingly, just as they were reflections of, and influences on his practice in unified field theory. The re-shaped methodological beliefs were readily invoked to justify the further mathematization of his research, and its increasing alienation from the realm of experience.

The semivector episode showed that Einstein's methodological beliefs were explicitly put to use when he was trying to make and motivate choices in his actual research. Semivectors appeared to deliver a most appealing result – the unified description of electrons and protons – and they figured prominently in the Spencer lecture. Soon it became clear, however, that only a Pyrrhic victory had been won; but Einstein, of course, did not retract his words spoken in Oxford on the method of theoretical physics.

From the lecture in Oxford, the Solovine schema and the work on unified field theory, a view emanates of what the older Einstein's ideal theoretical description of nature looked like, and how such a description should be formed. Experience should tickle the primarily mathematical intuition into forming a natural set of axioms; one may think of "natural" in the sense of symmetry principles and the like, but Einstein did not elaborate. Deduction from these fundamental axioms next should give the laws that relate back to experience in a more direct way. More inductive approaches, on the other hand, usually fail to produce true understanding and at best give mere phenomenological descriptions. Classical field theories retained the familiar notions of causality and the mathematics from relativity. These theories were to be put in their logically simplest form, which meant that the forces ought to be unified and that particles were to follow as solutions from the field equations. In this way one would arrive at a complete description of reality.

Einstein's treatment of the work of Klein exemplifies how he assessed the epistemic status of typical quantum relations. In a way one could say that Klein had achieved a unified structure that related the discreteness of charge to the inobservability of Kaluza's fifth dimension. A pivotal element in his reasoning was the de Broglie relation. Even though Einstein did consider a compactification of the fifth dimension and wanted to explain the quantized nature of the electric charge, he did not repeat Klein's argument. We believe that this was because he did not want to accept the de Broglie relation at the beginning of his theorizing. Rather, the de Broglie relation and the quantum of charge were to follow as Solovine S and E from unified field axioms A. This was exemplary of Einstein's attitude to the whole quantum theory: he did not deny its validity, but firmly believed that it was at best a mere phenomenological theory.

In the early 1920s Einstein had engrained himself in a different practice in physics than the quantum theorists; atom theorists studied different problems and took a different approach to theory construction. When this approach produced quantum mechanics, Einstein maintained that another kind of description and another kind of explanation can and ought to be found for atomic phenomena, to be created by a different method. His objections to quantum theory have usually been identified as largely philosophical. In our perspective these philosophical objections have acquired a certain circumstantial element; in part they were arguments looked for and found to give more epistemological capital to Einstein's own positive research program in mathematical unification.

Einstein's alternative, in the end, would really have been another generalized relativity theory. His words to de Broglie from 1953 – he looked like an "ostrich that keeps his head buried in relativistic sand" – suggest that the formulation of general relativity may have been such a shaping process that its effect can be compared to that of a physicist's training in a particular tradition of theory and problems. Andrew Warwick has recently shown – for the case of British theorists trained in the late Victorian era who did not appreciate the special theory of relativity[1] – that such an education can lead to a predisposition only to recognize arguments as convincing when they can be put in a form that is familiar from the perspective of the problems, themata and familiar approaches of a scholar's training.[2] Similarly, then, Einstein would have wanted to capture the atomistic character of nature with a theory that

---

[1] See Warwick (2003).

[2] Warwick's perspective shows similarities to elements of the historiographical approach of Jürgen Renn's group that may be relevant to point out here (even if there are also differences in focus, as Warwick has turned to didactic settings on the one hand, and Renn to cognitive structures on the other). It is namely natural to conceive of Renn's "mental models" (that is, knowledge structures built from resources of shared knowledge available at a given time; see Büttner *et al.* (2003), p. 45, Renn (2005a), p. 54) as the result of intensive training, or some other similarly formative experience.

retained the concepts and mathematics of general relativity, and foremost, he wanted to find his novel theory in a manner that resembled his warped recollection of the strong experience with the general theory of relativity; thus, he expected to find the theory using his idealized method. Quantum mechanics deviated too markedly from relativity in these respects and this prompted his negative attitude; it made him decide that his alternative attempt at theory construction should not yet end.

But this is only one side, since, as we have argued, the reverse holds true too: not only was Einstein directed to unified fields because of the experience and success of general relativity, but the recollection of the discovery of that theory was also shaped by his unified field theory attempts. He further employed his recollection to justify these attempts, just as his methodology was presented to enhance their profile. Initially, at the time of the Bad Nauheim debate or Einstein's 1923 programmatic paper on field theory and quanta, the unification program could claim to be an ambitious but not altogether unreasonable alternative to the problematic old quantum theory. Following the successes of Heisenberg and Schrödinger, and increasingly during the 1930s, this position became hard to maintain, and it is thus understandable that Einstein started referencing his method and the success of general relativity to try to regain some trust in, or even respectability for, his efforts.

Twentieth century physicists who worked on the quantum field description of elementary particles usually expressed themselves critical of the unified field theory program. For instance, Abraham Pais, who before he wrote his biography of Einstein had made a career in particle physics, held that "the time for unification had not yet come. [...] There are other forces in nature than gravitation and electromagnetism," a fact that Einstein, according to Pais, could, or rather, should have taken note of.[3] J. Robert Oppenheimer, Einstein's former director at the Institute for Advanced Study, chose more direct, harsher, words in an obituary, delivered at a Unesco colloquium in Paris in 1965: Einstein's last twenty-five years had been a "failure," and his attempt at unification a "hopelessly limited and historically rather accidentally conditioned approach." He had "lost contact with the profession of physics." In private, Oppenheimer would add that Einstein had been "wasting his time"; when he was still a young man, in 1935, Oppenheimer already deemed Einstein "completely cuckoo."[4]

These judgments suffer, however, from their own historical contingency, and were likely in considerable measure due to frustration with Einstein's opposition to quantum theory. Historians, who in any case usually shy away from passing

---

[3] See Pais (1982), p. 350; for Pais's critical attitude, see also p. 325.
[4] Oppenheimer as quoted in Schweber (2008), on pp. 279 and 276, respectively. "Cuckoo": J. Robert Oppenheimer to his brother Frank Oppenheimer, 11 January 1935, cited in Bird and Sherwin (2006), p. 64.

judgment, have been milder: Vladimir Vizgin has for instance pointed out that the larger unified field theory program did engage in creative cross-overs with the development of quantum mechanics – one can think for instance of the quantum theorists adoption of Einstein's local tetrad field (a concept developed for his teleparallel construction). The study of unified field theories would have given insights that proved indispensable in the development of gauge theory in particular.[5]

The first generation of quantum theorists – Pauli, Born, and Oppenheimer should be included too – had largely followed Bohr's lead on interpretative issues; they did not care for the EPR argument and adhered to what became known as the Copenhagen interpretation. They continued to construct theories in ways different from Einstein, looking to experiment first and engaging the mathematics of operator algebras. Unified fields were regarded as an extension of Einstein's rejection of quantum theory; they constituted an entirely outdated approach to the problems of the quantum and were beyond the pale in addressing pressing issues in nuclear and particle physics.

Pauli kept pointing out to Einstein the shortcomings of his models, but to no apparent avail. As suggested earlier, he may have hoped eventually to draw Einstein into the quantum program, or at least to persuade him to pass a blessing. Einstein did not particularly solicit Pauli's opinion, nor really anyone else's, and he comes across as a highly independent and single-minded thinker. It appears that in our story only Paul Ehrenfest had some direct influence on him; Ehrenfest was of course much more than any other correspondent someone of the same generation – both were Jewish intellectuals from the German speaking world with still enough nineteenth century, with its search for a unified *Weltbild*, in their bones. Einstein followed Weyl's work, but certainly not to the degree that one can say that it caused him to alter his own line of thought. The same image of Einstein presents itself when we look at his collaborations with Walther Mayer, Valentin Bargmann and Peter Bergmann. Bergmann was of course too much a junior to expect him to exert any great influence on Einstein. Bargmann appears to have been handpicked predominantly, if not exclusively, because of his expertise in mathematics, and the same was certainly true for Mayer. Their correspondence with Einstein shows that physical issues were hardly discussed. Every now and then Einstein would patiently sit through Mayer's or Bargmann's physical reasonings, politely break off the discussion and then return to the mathematical problems that he intended to address and for which he felt that he could draw on their expertise.

---

[5] See Vizgin (1994), pp. 306–307. The physicists Norbert Straumann and Lochlainn O'Raifeartaigh have found Vizgin's observations evident in one of the first papers on unified field theories, i.e. Weyl's theory (Weyl, 1918): Weyl's "brilliant proposal contains the germs of all mathematical aspects of a non-Abelian gauge theory," as in O'Raifeartaigh and Straumann (2000), p. 1. On the adoption of local tetrad fields by Weyl and Eugene Wigner, see O'Raifeartaigh (1997), p. 112.

Einstein's attempts in field theory were often forgotten within a few years after they had been presented. Many were characterized by a lack of contact with the world of experience, and his attempts at particle solutions ultimately all failed. Most quantum theorists marvelled in disbelief at his steadfastness in resisting their successful theory, and the dogged pursuit of alternatives.

Einstein's relocation to America was not likely to improve the reception of his work in professional circles. American physics, long before 1933, had been dominated by empiricist and, foremost, pragmatic attitudes – ironically, Percy W. Bridgman had been particularly instrumental in disseminating the pragmatic outlook. When EPR was published in 1935, most American physicists seemed uninterested, and European refugee theorists who worked on more experimental problems had an easier time finding positions in the United States than their more philosophically minded colleagues.[6] The next generation of American physicists, whose relative numbers had increased so much that they came to dominate the field internationally, was educated in the context of World War II and the Cold War. The large majority were not interested in foundational issues and held interpretative debates in quantum theory to be disputes about "metaphysics" that had no bearing on physics proper.[7] Such sentiment, by implication, would of course easily be extended to Einstein's unified field theory program as well. It thus found itself in an increasingly isolated position.

Another reason that stacked the odds against a successful integration of the field theory program in American academia was directly related to the constraints that the Cold War placed on the physics curriculum: physics departments had to educate ever larger numbers of Ph.D. students to meet the demands of funding agencies with military agendas. Unified field theory research was neither thematically nor practically suited: the subject was far removed from all things nuclear, but perhaps even more importantly, it did not lend itself easily to graduate research and quick Ph.D.s. For this reason, the Berkeley physics department, the largest in the United States, advised one of its junior faculty members, who was pursuing Einstein's program, to look for a position elsewhere; his research was not useful for graduating expeditiously large numbers of students.[8]

Empiricism and pragmatism remained the dominant attitudes in the development of quantum field theory: theorists such as Richard Feynman and Hans Bethe worked towards "getting numbers out."[9] Due to the Cold War, funding for elementary particle physics was abundant; research in general relativity, however, was in

[6] On American theoretical physics in the period 1920–1950, see Schweber (1986); see also Frans van Lunteren, "Theoretische fysica als zelfstandige discipline" (typescript, 2008).

[7] See Freire (2005), pp. 23–24, quoting Fritz Bopp in 1957. Recently Lee Smolin has lamented the dominance of this attitude during his graduate education in the 1970s; see Smolin (2006), p. 312.

[8] See Kaiser (2002), in particular pp. 155–156.

[9] According to John A. Wheeler, as cited on p. 96 in Schweber (1986).

relative decline.[10] As scholars pursuing relativity judged Einstein's program less critically than most quantum field theorists, these general trends further boded ill for the overall assessment of Einstein's program.

The above illustrates once again that different groups in the physics community with different programmatic preferences and research practices arrived at different appreciations. Lately, some string theorists have claimed Einstein's unification attempts to be the vanguard efforts of a visionary forerunner. String theory also attempts a unification of all of nature's forces, and its methods are largely mathematical too – of the esthetic, highly paced and intuitive kind, rather than the slow moving, very precise variety. Similarities with Einstein's approaches are thus obvious; the theory, even though it developed out of the quantum field program, has finally also been challenged as lacking substantial contact with experience.

String theorist Brian Greene, in his best-selling book *The Elegant Universe*, has qualified Einstein's work as follows:

In Einstein's day, the strong and the weak forces had not yet been discovered, but he found the existence of even two distinct forces – gravity and electromagnetism – deeply troubling. Einstein did not accept that nature is founded on such an extravagant design. This launched his thirty-year voyage in search of the so-called *unified field theory* that he hoped would show that these two forces are really manifestations of one grand underlying principle. This quixotic quest isolated Einstein from the mainstream of physics [...].

Einstein was simply ahead of his time. More than half a century later, his dream of a unified theory has become the Holy Grail of modern physics. And a sizable part of the physics and mathematics community is becoming increasingly convinced that string theory may provide the answer.[11]

The notion that Einstein was "ahead of his time" stands in stark contrast to the strong words of Oppenheimer. Pais, like Greene, had felt that Einstein had started too early with his work on unification, but as said, he pointed out rather critically that, in fact, Einstein could and should have taken note of nuclear forces as they were being unearthed while he still only concerned himself with gravity and electromagnetism. Clearly, all these judgments, even Greene's portrayal of Einstein's work, depend on their own particular historical contingency.

Some may disagree, however, and suggest that we use Einstein's unification attempts as a negative example to assess current practice in theory construction.[12] But one swallow does not make a summer: for such discussions, one may wish to engage more cases than just Einstein's example. Besides, along with the similar roles of maxims like "mathematical naturalness" or "logical simplicity" in the string and unified field approaches, there are also at least as many differences: not only did

[10] On relativity, see Eisenstaedt (1989b).
[11] As in Greene (1999), p. 15.
[12] For an example that comes close to this way of approaching the issue, see Woit (2006), p. 187.

Einstein ignore nuclear forces, he also of course did not quantize his theories. The latter is a major difference with most, if not all, modern unification attempts. This last point, however, also makes clear a larger problem with the kind of comparison suggested: such comparisons presume that a historian can strip Einstein, as well as modern theorists, from their relevant contexts; they imply that one can abstract scholars away from the precise questions and concepts that are essential to their way of viewing their problem. This kind of assessment is of course alien to the historian's enterprise; as said earlier, historians do not strive to find a recipe for a timeless method for doing theory, and neither do they posses a golden standard with which to compare Einstein's efforts or those of string theorists. It is up to today's physicists to decide to what degree Einstein's problems and approaches are still relevant for their own efforts; the historian or philosopher should not assume an arbitrating but instead a describing and interpreting role.

Yet we never intended to address such wide panoramas. We only wanted to understand Einstein's historical development and to see how he came to his dissenting position. An emotionally defining moment was the discovery of general relativity. It was instrumental in locking him, eventually, in a belief in his idealized method and the pursuit of unified field theories. Pursuing unified field theories and resisting quantum mechanics, validated by a one-sided recollection of the experience of relativity, further meant that familiar field concepts could be retained. When creating new theories, a personal sense of mathematical naturalness in effect came before experiment, and unification became a near synonym of explanation. The quantum ultimately had to be subsumed to a theory of the "total" field – ultimately, Einstein was certain that the successful method of general relativity had to be repeated.

References

Abiko S. (2000). Einstein's Kyoto address: "How I created the theory of relativity." *Historical Studies in the Physical and Biological Sciences* **31**, 1–35.

Abraham M. (1912a). Zur Theorie der Gravitation. *Physikalische Zeitschrift* **13**, 1–4.

Abraham M. (1912b). Relativität und Gravitation. Erwiderung auf eine Bemerkung des Hrn. A. Einstein. *Annalen der Physik* **38**, 1056–1058.

Abraham M. (1912c). Nochmals Relativität und Gravitation. Bemerkungen zu A. Einsteins Erwiderung. *Annalen der Physik* **39**, 444–448.

Abraham M. (1914). Die neue Mechanik. *Scientia* **15**, 8–27.

Adam A. (2000). Farewell to certitude: Einstein's novelty on induction and deduction, fallibilism. *Journal for General Philosophy of Science* **31**, 19–37.

Atkinson R. (1926). Über Interferenz von Kanalstrahlenlicht. *Die Naturwissenschaften* **14**, 599–600.

Bargmann V. (1934). Über den Zusammenhang zwischen Semivektoren und Spinoren und die Reduktion der Diracgleichungen für Semivektoren. *Helvetica Physica Acta* **7**, 57–82.

Bargmann V. (1979). Erinnerungen eines Assistenten Einsteins. *Vierteljahrschrift der Naturforschenden Gesellschaft in Zürich* **124**, 39–44.

Bell J. S. (1987). *Speakable and Unspeakable in Quantum Mechanics*. Cambridge: Cambridge University Press.

Beller M. (1999). *Quantum Dialogue*. Chicago, IL: University of Chicago Press.

Beller M. (2000). Kant's impact on Einstein's thought. In *Einstein – The Formative Years 1879–1909*, D. A. Howard and J. Stachel, eds. Boston, MA: Birkhäuser, 83–106.

Belousek D. W. (1996). Einstein's 1927 unpublished hidden-variable theory: Its background, context and significance. *Studies in History and Philosophy of Modern Physics* **27**, 437–461.

Bergia S. (1993). Attempts at unified field theories (1919–1955). Alleged failure and intrinsic validation/refutation criteria. In *The Attraction of Gravitation. New Studies in the History of General Relativity*, J. Earman, M. Janssen, and J. D. Norton, eds. Boston, MA: Birkhäuser, 274–307.

Bergmann P. G. (1942). *Introduction to the Theory of Relativity*. New York: Prentice-Hall.

Bergmann P. G. (1949). Non-linear field theories. *Physical Review* **75**, 680–685.

Bergmann P. G. (1992). Quantization of the gravitational field, 1930–1988. In *Studies in the History of General Relativity*, J. Eisenstaedt and A. J. Kox, eds. Boston, MA: Birkhäuser, 364–366.

Bergmann P. G., Brunings J. H. (1949). Non-linear field theories II: Canonical equations and quantization. *Reviews of Modern Physics* **21**, 480–487.

Bergmann P. G., Penfield R., Schiller R., Zatzkis H. (1950). The Hamiltonian of the general theory of relativity with electromagnetic field. *Physical Review* **80**, 81–88.

Bird K., Sherwin M. J. (2006). *American Prometheus. The Triumph and Tragedy of J. Robert Oppenheimer*. New York: Vintage.

Blaton J. (1935). Quaternionen, Semivektoren und Spinoren. *Zeitschrift für Physik* **95**, 337–354.

Bohr N. (1913). On the constitution of atoms and molecules. Part I. *Philosophical Magazine* **26**, 1–25.

Bohr N. (1922). *Drei Aufsätze über Spektren und Atombau*. Braunschweig: Vieweg.

Bohr N. (1949). Discussion with Einstein on epistemological problems in atomic physics. In *Albert Einstein: Philosopher-Scientist*, P. A. Schilpp, ed. La Salle, IL: Open Court, 199–241.

Bridgman P. W. (1949). Einstein's theories and the operational point of view. In *Albert Einstein: Philosopher-Scientist*, P. A. Schilpp, ed. La Salle, IL: Open Court, 335–354.

Büchner L. (1888). *Kraft und Stoff, oder Grundzüge der natürlichen Weltordnung, nebst einer darauf gebauten Moral oder Sittenlehre*, 16 edn. Leipzig: Theodor Thomas.

Buchwald D. K., Schulmann R., Illy J., Kennefick D. J., Sauer T., Eds. (2004). *The Collected Papers of Albert Einstein*. Vol. 9, Correspondence 1919–April 1920. Princeton, NJ: Princeton University Press.

Buchwald D. K., Rosenkranz Z., Sauer T., Illy J., Holmes V. I., Eds. (2009). *The Collected Papers of Albert Einstein*. Vol. 12, Correspondence 1921. Princeton, NJ: Princeton University Press.

Büttner J., Renn J., Schemmel M. (2003). Exploring the limits of classical physics: Planck, Einstein and the structure of a scientific revolution. *Studies in History and Philosophy of Modern Physics* **34**, 37–59.

Cartan É. (1923). Sur les variétés à connexion affine et la théorie de la rélativité généralisée. *Annales de l'École Normale Supérieure* **40**, 325–412.

Cartan É. (1966). *The Theory of Spinors*. Cambridge, MA: MIT Press.

Casimir H. B. G. (1983). *Het toeval van de werkelijkheid. Een halve eeuw natuurkunde*. Amsterdam: Meulenhoff.

Cassidy D. C. (2008). Re-examining the crisis in quantum theory, part I: Spectroscopy. In *HQ1: Conference on the History of Quantum Physics*, C. Joas, C. Lehner and J. Renn, eds. Vol. 1. Berlin: Max Planck Institute for the History of Science, Preprint 350, 105–115.

Cattani C., de Maria M. (1989). The 1915 epistolary controversy between Einstein and Tullio Levi-Civita. In *Einstein and the History of General Relativity*, D. A. Howard and J. Stachel, eds. Boston, MA: Birkhäuser, 175–200.

Chandrasekharan K., Ed. (1968). *Hermann Weyl. Gesammelte Abhandlungen*. Vol. III. Berlin: Springer.

Cohen R. S. (1970). Ernst Mach: Physics, perception and the philosophy of science. In *Ernst Mach: Physicist and Philosopher*, R. S. Cohen and R. J. Seeger, eds. Dordrecht: Reidel, 126–164.

Corry L. (1998). The influence of David Hilbert and Hermann Minkowski on Einstein's views over the interrelation between physics and mathematics. *Endeavour* **22**, 95–97.

Corry L., Renn J., Stachel J. (1997). Belated decision in the Hilbert–Einstein priority dispute. *Science* **278**, 1270–1273.

Cushing J. T. (1994). *Quantum Mechanics: Historical Contingency and the Copenhagen Hegemony*. Chicago, IL: University of Chicago Press.

Davisson C., Germer L. (1927). Diffraction of electrons by a crystal of nickel. *Physical Review* **30**, 705–740.

Debever R., Ed. (1979). *Élie Cartan–Albert Einstein. Letters on Absolute Parallelism 1929–1932.* Princeton, NJ: Princeton University Press.

de Regt H. (1997). Erwin Schrödinger, Anschaulichkeit, and quantum theory. *Studies in History and Philosophy of Modern Physics* **28**, 461–481.

de Regt H. (2001). Spacetime visualisation and the intelligibility of physical theories. *Studies in History and Philosophy of Modern Physics* **32**, 243–265.

Dieks D. (1994). The scientific view of the world: Introduction. In *Physics and Our View of the World*, J. Hilgevoord, ed. Cambridge: Cambridge University Press, 61–78.

Dieks D. (2010). The adolescence of relativity: Einstein, Minkowski, and the philosophy of space and time. In *Minkowski Spacetime: One Hundred Years Later*, V. Petkov, ed. New York: Springer, 225–247.

Dieks D., Lam S. (2008). Complementarity in the Einstein–Bohr photon box. *American Journal of Physics* **76**, 838–842.

Dirac P. A. M. (1928a). The quantum theory of the electron. *Proceedings of the Royal Society of London* **A117**, 610–624.

Dirac P. A. M. (1928b). The quantum theory of the electron. Part II. *Proceedings of the Royal Society of London* **A118**, 351–361.

Dirac P. A. M. (1930a). *The Principles of Quantum Mechanics*. Oxford: Oxford University Press.

Dirac P. A. M. (1930b). A theory of electrons and protons. *Proceedings of the Royal Society of London* **A126**, 360–365.

Duff M. J., Nilsson B. E. W., Pope C. N. (1986). Kaluza–Klein supergravity. *Physics Reports* **130**, 1–142.

Duncan A., Janssen M. (2007). On the verge of Umdeutung in Minnesota: Van Vleck and the correspondence principle. Parts I and II. *Archive for History of Exact Sciences* **61**, 553–624, 625–671.

Earman J., Eisenstaedt J. (1999). Einstein and singularities. *Studies in History and Philosophy of Modern Physics* **30**, 185–235.

Earman J., Glymour C. (1980). Relativity and eclipses: The British eclipse expeditions of 1919 and their predecessors. *Historical Studies in the Physical Sciences* **11**, 49–85.

Earman J., Janssen M. (1993). Einstein's explanation of the motion of Mercury's perihelion. In *The Attraction of Gravitation. New Studies in the History of General Relativity*, J. Earman, M. Janssen and J. D. Norton, eds. Boston, MA: Birkhäuser, 129–172.

Earman J., Norton J. D. (1987). What price spacetime substantivalism? The hole story. *British Journal for the Philosophy of Science* **38**, 515–525.

Eckert M., Märker K., Eds. (2004). *Arnold Sommerfeld. Wissenschaftlicher Briefwechsel.* Vol. 2, 1919–1951. Diepholz: GNT-Verlag.

Eddington A. S. (1933). *The Expanding Universe*. Cambridge: Cambridge University Press.

Ehrenfest P. (1909). Gleichförmige Rotation starrer Körper und Relativitätstheorie. *Physikalische Zeitschrift* **10**, 918.

Ehrenfest P. (1932). Einige die Quantenmechanik betreffende Erkundigungsfragen. *Zeitschrift für Physik* **78**, 555–559.

Einstein A. (1905a). Über einen die Erzeugung und Verwandlung des Lichtes betreffenden heuristischen Gesichtspunkt. *Annalen der Physik* **17**, 132–148.

Einstein A. (1905b). Über die von der molekularkinetischen Theorie der Wärme geforderte Bewegung von in ruhenden Flüssigkeiten suspendierten Teilchen. *Annalen der Physik* **17**, 549–560.

194   *References*

Einstein A. (1905c). Zur Elektrodynamik bewegter Körper. *Annalen der Physik* **17**, 891–921.

Einstein A. (1907a). Über die Gültigkeitsgrenze des Satzes vom thermodynamischen Gleichgewicht und über die Möglichkeit einer neuen Bestimmung der Elementarquanta. *Annalen der Physik* **22**, 569–572.

Einstein A. (1907b). Über das Relativitätsprinzip und die aus demselben gezogenen Folgerungen. *Jahrbuch der Radioaktivität und Elektronik* **4**, 411–462.

Einstein A. (1909a). Zum gegenwärtigen Stand des Strahlungsproblems. *Physikalische Zeitschrift* **10**, 185–193.

Einstein A. (1909b). Über die Entwickelung unserer Anschauungen über das Wesen und die Konstitution der Strahlung. *Verhandlungen der Deutschen Physikalischen Gesellschaft* **7**, 482–500.

Einstein A. (1911). Über den Einfluss der Schwerkraft auf die Ausbreitung des Lichtes. *Annalen der Physik* **35**, 898–908.

Einstein A. (1912a). Lichtgeschwindigkeit und Statik des Gravitationsfeldes. *Annalen der Physik* **38**, 355–369.

Einstein A. (1912b). Zur Theorie des statischen Gravitationsfeldes. *Annalen der Physik* **38**, 443–458.

Einstein A. (1912c). Relativität und Gravitation. Erwiderung auf eine Bemerkung von M. Abraham. *Annalen der Physik* **38**, 1059–1064.

Einstein A. (1912d). Bemerkung zu Abrahams vorangehender Auseinandersetzung "Nochmals Relativität und Gravitation." *Annalen der Physik* **39**, 704.

Einstein A. (1914a). Prinzipielles zur verallgemeinerten Relativitätstheorie und Gravitationstheorie. *Physikalische Zeitschrift* **15**, 176–180.

Einstein A. (1914b). Antrittsrede. *Sitzungsberichte der Königlich Preußischen Akademie der Wissenschaften*, 739–742.

Einstein A. (1914c). Die formale Grundlage der allgemeinen Relativitätstheorie. *Sitzungsberichte der Königlich Preußischen Akademie der Wissenschaften*, 1030–1085.

Einstein A. (1915a). Zur allgemeinen Relativitätstheorie. *Sitzungsberichte der Königlich Preußischen Akademie der Wissenschaften*, 778–786.

Einstein A. (1915b). Zur allgemeinen Relativitätstheorie. (Nachtrag). *Sitzungsberichte der Königlich Preußischen Akademie der Wissenschaften*, 799–801.

Einstein A. (1915c). Erklärung der Perihelbewegung des Merkur aus der allgemeinen Relativitätstheorie. *Sitzungsberichte der Königlich Preußischen Akademie der Wissenschaften*, 831–839.

Einstein A. (1915d). Die Feldgleichungen der Gravitation. *Sitzungsberichte der Königlich Preußischen Akademie der Wissenschaften*, 844–847.

Einstein A. (1916a). Ernst Mach. *Physikalische Zeitschrift* **17**, 101–104.

Einstein A. (1916b). Zur Quantentheorie der Strahlung. *Physikalische Gesellschaft Zürich. Mitteilungen* **18**, 47–62.

Einstein A. (1918). Motive des Forschens. In *Zu Max Plancks sechzigstem Geburtstag. Ansprachen, gehalten am 26. April 1918 in der Deutschen Physikalischen Gesellschaft von E. Warburg, M. v. Laue, A. Sommerfeld und A. Einstein.* Karlsruhe: C. F. Müllersche Hofbuchhandlung, 29–32.

Einstein A. (1919a). Spielen Gravitationsfelder im Aufbau der materiellen Elementarteilchen eine wesentliche Rolle? *Sitzungsberichte der Königlich Preußischen Akademie der Wissenschaften*, 349–356.

Einstein A. (1919b). Time, space, and gravitation. *The Times*, 28 November, 13–14.

Einstein A. (1919c). Induktion und Deduktion in der Physik. *Berliner Tageblatt*, 25 December, Morning edition, 4. Beiblatt, 1.

Einstein A. (1921a). *Geometrie und Erfahrung*. Berlin: Springer.

Einstein A. (1921b). Über ein den Elementarprozess der Lichtemission betreffendes Experiment. *Sitzungsberichte der Königlich Preußischen Akademie der Wissenschaften*, 882–883.

Einstein A. (1922). Theoretische Bemerkungen zur Supraleitung der Metalle. In *Het Natuurkundig Laboratorium der Rijksuniversiteit te Leiden in de jaren 1904–1922: gedenkboek aangeboden aan H. Kamerlingh Onnes bij gelegenheid van zijn veertigjarig professoraat op 11 november 1922*, C. A. Crommelin, ed. Leiden: IJdo, 429–435.

Einstein A. (1923a). Grundgedanken und Probleme der Relativitätstheorie. In *Nobelstiftelsen, Les Prix Nobel en 1921–1922*. Stockholm: Imprimerie Royale, 1–10.

Einstein A. (1923b). Bietet die Feldtheorie Möglichkeiten für die Lösung des Quantenproblems? *Sitzungsberichte der Preußischen Akademie der Wissenschaften*, 359–364.

Einstein A. (1924). Quantentheorie des einatomigen idealen Gases. *Sitzungsberichte der Preußischen Akademie der Wissenschaften*, 261–267.

Einstein A. (1925a). Quantentheorie des einatomigen idealen Gases. 2. Abhandlung. *Sitzungsberichte der Preußischen Akademie der Wissenschaften*, 3–14.

Einstein A. (1925b). Nichteuklidische Geometrie und Physik. *Die neue Rundschau*, January, 16–20.

Einstein A. (1925c). Quantentheorie des idealen Gases. *Sitzungsberichte der Preußischen Akademie der Wissenschaften*, 18–25.

Einstein A. (1926a). Vorschlag zu einem die Natur des elementaren Strahlungs-Emissionsprozesses betreffenden Experiment. *Die Naturwissenschaften* **14**, 300–301.

Einstein A. (1926b). Über die Interferenzeigenschaften des durch Kanalstrahlen emittierten Lichtes. *Sitzungsberichte der Preußischen Akademie der Wissenschaften*, 334–340.

Einstein A. (1927). Zu Kaluzas Theorie des Zusammenhanges von Gravitation und Elektrizität. Erste und zweite Mitteilung. *Sitzungsberichte der Preußischen Akademie der Wissenschaften*, 23–25, 26–30.

Einstein A. (1928a). Riemann-Geometrie mit Aufrechterhaltung des Begriffes des Fernparallelismus. *Sitzungsberichte der Preußischen Akademie der Wissenschaften*, 217–221.

Einstein A. (1928b). Neue Möglichkeit für eine einheitliche Feldtheorie von Gravitation und Elektrizität. *Sitzungsberichte der Preußischen Akademie der Wissenschaften*, 224–227.

Einstein A. (1928c). À propos de 'La déduction relativiste' de M. Emile Meyerson. *Revue Philosophique de la France et de l'Étranger* **105**, 161–166.

Einstein A. (1929). Über den gegenwärtigen Stand der Feldtheorie. In *Festschrift zum 70. Geburtstag von Prof. Dr. A. Stodola*, E. Honegger, ed. Zurich: Füssli, 126–132.

Einstein A. (1931). Maxwell's influence on the development of the conception of physical reality. In *James Clerk Maxwell: A Commemoration Volume, 1831–1931*. Cambridge: Cambridge University Press, 66–73.

Einstein A. (1932). Der gegenwärtige Stand der Relativitätstheorie. *Die Quelle* **82**, 440–442.

Einstein A. (1933a). *On the Method of Theoretical Physics*. Oxford: Clarendon Press.

Einstein A. (1933b). *The Origins of the General Theory of Relativity*. Glasgow: Jackson.

Einstein A. (1936a). Physik und Realität. *Journal of the Franklin Institute* **221**, 313–347.

Einstein A. (1936b). Lens-like action of a star by deviation of light in the gravitational field. *Science* **84**, 506–507.

Einstein A. (1939). On a stationary system with spherical symmetry consisting of many gravitating masses. *Annals of Mathematics* **40**, 922–936.

Einstein A. (1940). Considerations concerning the fundaments of theoretical physics. *Science* **91**, 487–492.

Einstein A. (1941). Demonstration of the non-existence of gravitational fields with a non-vanishing total mass free of singularities. *Revista Universidad Nacional de Tucuman* **A2**, 11–16.

Einstein A. (1944). Remarks on Bertrand Russell's theory of knowledge. In *The Philosophy of Bertrand Russell*, P. A. Schilpp, ed. Evanston, IL: Northwestern University Press, 278–291.

Einstein A. (1949a). Autobiographisches. In *Albert Einstein: Philosopher-Scientist*, P. A. Schilpp, ed. La Salle, IL: Open Court, 1–94.

Einstein A. (1949b). Remarks concerning the essays brought together in this co-operative volume. In *Albert Einstein: Philosopher-Scientist*, P. A. Schilpp, ed. La Salle, IL: Open Court, 663–688.

Einstein A. (1950). On the generalized theory of gravitation. *Scientific American* **182**, 13–17.

Einstein A. (1955). *The Meaning of Relativity*, 5 edn. Princeton, NJ: Princeton University Press.

Einstein A. (1956). *Lettres à Maurice Solovine*. Paris: Gauthier-Villars.

Einstein A. (1994). *Ideas and Opinions*. New York: The Modern Library.

Einstein A. (2008). Unpublished opening lecture for the course on the theory of relativity in Argentina, 1925. *Science in Context* **21**, 451–459.

Einstein A., Bargmann V. (1944). Bivector fields, I. *Annals of Mathematics* **45**, 1–14.

Einstein A., Bergmann P. G. (1938). On a generalisation of Kaluza's theory of electricity. *Annals of Mathematics* **39**, 683–701.

Einstein A., Besso M. (1972). *Correspondance, 1903–1955*. Paris: Hermann.

Einstein A., de Haas W. (1915). Experimenteller Nachweis der Ampèreschen Molekularströme. *Verhandlungen der Deutschen Physikalischen Gesellschaft* **17**, 152–170.

Einstein A., de Haas W. (1915–16). Experimental proof of the existence of Ampère's molecular currents. *Proceedings of the Section of Sciences of the Koninklijke Akademie van Wetenschappen* **18**, 696–711.

Einstein A., Ehrenfest P. (1922). Quantentheoretische Bemerkungen zum Experiment von Stern und Gerlach. *Zeitschrift für Physik* **11**, 31–34.

Einstein A., Ehrenfest P. (1923). Zur Quantentheorie des Strahlungsgleichgewichts. *Zeitschrift für Physik* **19**, 301–306.

Einstein A., Fokker A. (1914). Die Nordströmsche Gravitationstheorie vom Standpunkt des absoluten Differentialkalküls. *Annalen der Physik* **44**, 321–328.

Einstein A., Grommer J. (1923). Beweis der Nichtexistenz eines überall regulären zentrisch symmetrischen Feldes nach der Feld-Theorie von Th. Kaluza. *Scripta Universitatis atque Bibliothecae Hierosolymitanarum: Mathematica et Physica* **1**, 1–5.

Einstein A., Grossmann M. (1913). *Entwurf einer verallgemeinerten Relativitätstheorie und einer Theorie der Gravitation*. Leipzig: Teubner.

Einstein A., Grossmann M. (1914). Kovarianzeigenschaften der Feldgleichungen der auf die verallgemeinerte Relativitätstheorie gegründeten Gravitationstheorie. *Zeitschrift für Mathematik und Physik* **63**, 215–225.

Einstein A., Mayer W. (1930). Zwei strenge statische Lösungen der Feldgleichungen der einheitlichen Feldtheorie. *Sitzungsberichte der Preußischen Akademie der Wissenschaften*, 110–120.

Einstein A., Mayer W. (1931). Einheitliche Theorie von Gravitation und Elektrizität. *Sitzungsberichte der Preußischen Akademie der Wissenschaften*, 541–557.

Einstein A., Mayer W. (1932a). Einheitliche Theorie von Gravitation und Elektri- zität. (Zweite Abhandlung). *Sitzungsberichte der Preußischen Akademie der Wis- senschaften*, 130–137.

Einstein A., Mayer W. (1932b). Semi-Vektoren und Spinoren. *Sitzungsberichte der Preußischen Akademie der Wissenschaften*, 522–550.

Einstein A., Mayer W. (1933a). Die Dirac Gleichungen für Semi-Vektoren. *Proceedings of the Section of Sciences of the Koninklijke Akademie van Wetenschappen* **36**, 497–516.

Einstein A., Mayer W. (1933b). Spaltung der natürlichsten Feldgleichungen für Semi- Vektoren in Spinor-Gleichungen vom Dirac'schen Typus. *Proceedings of the Section of Sciences of the Koninklijke Akademie van Wetenschappen* **36**, 615–619.

Einstein A., Mayer W. (1934). Darstellung der Semi-Vektoren als gewönliche Vektoren von besonderem Differentiations Charakter. *Annals of Mathematics* **35**, 104–110.

Einstein A., Mühsam H. (1923). Experimentelle Bestimmung der Kanalweite von Filtern. *Deutsche medizinische Wochenschrift* **49**, 1012–1013.

Einstein A., Pauli W. (1943). On the non-existence of regular stationary solutions of relativistic field equations. *Annals of Mathematics* **44**, 131–137.

Einstein A., Rosen N. (1935). The particle problem in the general theory of relativity. *Physical Review* **48**, 73–77.

Einstein A., Rosen N. (1936). On gravitational waves. *Journal of the Franklin Institute* **223**, 43–54.

Einstein A., Sommerfeld A. (1968). *Albert Einstein–Arnold Sommerfeld, Briefwechsel.* Basel: Schwabe.

Einstein A., Podolsky B., Rosen N. (1935). Can quantum mechanical description of physical reality be considered complete? *Physical Review* **47**, 777–780.

Einstein A., Infeld L., Hoffmann B. (1938). The gravitational equations and the problem of motion. *Annals of Mathematics* **39**, 65–100.

Einstein A., Bargmann V., Bergmann P. G. (1941). On the five-dimensional representation of gravitation and electricity. In *Theodore von Karman Anniversary Volume.* Pasadena, CA: California Institute of Technology, 212–225.

Einstein A., Lorentz H., Planck M., Schrödinger E. (1963). *Briefe zur Wellenmechanik.* Vienna: Springer.

Einstein A., Born H., Born M. (1969). *Briefwechsel 1916–1955.* Munich: Nymphenburger Verlagshandlung.

Eisenstaedt J. (1982). Histoire et singularités de la solution de Schwarzschild (1915–1923). *Archive for History of Exact Sciences* **27**, 157–198.

Eisenstaedt J. (1987). Trajectoires et impasses de la solution de Schwarzschild. *Archive for History of Exact Sciences* **37**, 275–357.

Eisenstaedt J. (1989a). The early interpretation of the Schwarzschild solution. In *Einstein and the History of General Relativity*, D. A. Howard and J. Stachel, eds. Boston, MA: Birkhäuser, 213–233.

Eisenstaedt J. (1989b). The low water mark of general relativity, 1925–1955. In *Einstein and the History of General Relativity*, D. A. Howard and J. Stachel, eds. Boston, MA: Birkhäuser, 277–292.

Eisenstaedt J. (1993). Lemaître and the Schwarzschild solution. In *The Attraction of Grav- itation. New Studies in the History of General Relativity*, J. Earman, M. Janssen and J. D. Norton, eds. Boston, MA: Birkhäuser, 353–389.

Fine A. (1996). *The Shaky Game. Einstein, Realism and the Quantum Theory*, 2 edn. Chicago, IL: University of Chicago Press.

Fock V. (1926). Über die invariante Form der Wellen- und der Bewegungsgleichungen für einen geladenen Massenpunkt. *Zeitschrift für Physik* **39**, 226–232.

Fölsing A. (1998). *Albert Einstein. A Biography*. New York: Penguin.

Forman P. (1970). Alfred Landé and the anomalous Zeeman effect, 1919–1921. *Historical Studies in the Physical Sciences* **2**, 153–261.

Forman P. (1971). Weimar culture, causality, and quantum theory: Adaptation by German physicists and mathematicians to a hostile environment. *Historical Studies in the Physical Sciences* **3**, 1–115.

Freire O. (2005). Science and exile: David Bohm, the cold war, and a new interpretation of quantum mechanics. *Historical Studies in the Physical and Biological Sciences* **36**, 1–34.

French A. P. (1999). The strange case of Emil Rupp. *Physics in Perspective* **1**, 3–21.

Friedman M. (1974). Explanation and scientific understanding. *Journal of Philosophy* **71**, 5–19.

Friedman M. (1992). Causal laws and the foundations of natural science. In *The Cambridge Companion to Kant*, P. Guyer, ed. Cambridge: Cambridge University Press, 161–199.

Galison P. (1987). *How Experiments End*. Chicago, IL: University of Chicago Press.

Galvagno M., Giribet G. (2005). The particle problem in classical gravity: A historical note on 1941. *European Journal of Physics* **26**, S97–S110.

Gearheart C. A. (2002). Planck, the quantum, and the historians. *Physics in Perspective* **4**, 170–215.

Gerlach W., Rüchardt E. (1935). Über die Kohärenzlänge des von Kanalstrahlen emittierten Lichtes. *Annalen der Physik* **24**, 124–126.

Giulini D. (2008). Concepts of symmetry in the work of Wolfgang Pauli. In *Recasting Reality. Wolfgang Pauli's Philosophical Ideas and Contemporary Science*, H. Atmanspacher and H. Primas, eds. Berlin: Springer, 33–82.

Glymour C. (1999). Realism and the nature of theories. In *Introduction to the Philosophy of Science*. Indianapolis, IN: Hackett, 104–131.

Goenner H. (1993). The reaction to relativity theory I: The anti-Einstein campaign in Germany in 1920. *Science in Context* **6**, 107–133.

Goenner H. (2004). On the history of unified field theories. *Living Reviews in Relativity* **7**, http://www.livingreviews.org/lrr–2004–2. Downloaded on 11 May 2004.

Goenner H. (2005). Unified field theory: Early history and interplay between mathematics and physics. In *The Universe of General Relativity*, A. J. Kox and J. Eisenstaedt, eds. Boston, MA: Birkhäuser, 303–325.

Goenner H., Wünsch D. (2003). *Kaluza's and Klein's contributions to Kaluza–Klein theory*. Berlin: Max Planck Institute for the History of Science, Preprint 235.

Goldstein C., Ritter J. (2003). The varieties of unity: Sounding unified field theories 1920–1930. In *Revisiting the Foundations of Relativistic Physics. Festschrift in Honor of John Stachel*, A. Ashtekar, R. S. Cohen, D. A. Howard, J. Renn, S. Sarkar and A. Shimony, eds. Dordrecht: Springer, 93–149.

Greene B. (1999). *The Elegant Universe. Superstrings, Hidden Dimensions, and the Quest for the Ultimate Theory*. New York: W. W. Norton.

Gregory F. (2000). The mysteries and wonders of natural science: Aaron Bernstein's Naturwissenschaftliche Volksbücher and the adolescent Einstein. In *Einstein: The Formative Years, 1879–1909*, D. A. Howard and J. Stachel, eds. Boston, MA: Birkhäuser, 23–41.

Gross D. J., Perry M. J. (1983). Magnetic monopoles in Kaluza–Klein theories. *Nuclear Physics* **B226**, 29–48.

Hacohen M. H. (2000). *Karl Popper. The Formative Years, 1902–1945. Politics and Philosophy in Interwar Vienna*. Cambridge: Cambridge University Press.

Hadamard J. (1949). *An Essay on the Psychology of Invention in the Mathematical Field*. Princeton, NJ: Princeton University Press.

Halpern P. (2004). *The Great Beyond. Higher Dimensions, Parallel Universes, and the Extraordinary Search for a Theory of Everything.* Hoboken, NJ: John Wiley & Sons.

Halpern P. (2005). Peter Bergmann: The education of a physicist. *Physics in Perspective* **7**, 390–403.

Halpern P. (2007). Klein, Einstein, and five-dimensional unification. *Physics in Perspective* **9**, 390–405.

Heilbron J. L., Kuhn T. S. (1969). The genesis of the Bohr atom. *Historical Studies in the Physical Sciences* **1**, 211–290.

Heisenberg W. (1925). Über quantentheoretische Umdeutung kinematischer und mechanischer Beziehungen. *Zeitschrift für Physik* **33**, 879–893.

Heisenberg W. (1969). *Der Teil und das Ganze. Gespräche im Umkreis der Atomphysik.* Munich: R. Piper.

Heisenberg W. (1989a). The beginnings of quantum mechanics in Göttingen. In *Encounters with Einstein and Other Essays on People, Places, and Particles*. Princeton, NJ: Princeton University Press, 37–55.

Heisenberg W. (1989b). Encounters and conversations with Albert Einstein. In *Encounters with Einstein and Other Essays on People, Places, and Particles*. Princeton, NJ: Princeton University Press, 107–122.

Hentschel K. (1986). Die Korrespondenz Einstein–Schlick: Zum Verhältnis der Physik zur Philosophie. *Annals of Science* **43**, 475–488.

Hentschel K. (1992). Einstein's attitude towards experiments: Testing relativity theory 1907–1927. *Studies in History and Philosophy of Science* **23**, 593–624.

Hermann A., von Meyenn K., Weisskopf V. F., Eds. (1979). *Wolfgang Pauli. Wissenschaftlicher Briefwechsel mit Bohr, Einstein, Heisenberg u.a.* Vol. I, 1919–1929. New York: Springer.

Hilbert D. (1915). Die Grundlagen der Physik. (Erste Mitteilung). *Nachrichten von der Königlichen Gesellschaft der Wissenschaften zu Göttingen*, 395–407.

Holton G. (1968). Mach, Einstein, and the search for reality. *Daedalus* **97**, 636–673.

Holton G. (1969). Einstein, Michelson, and the 'crucial' experiment. *Isis* **60**, 133–197.

Holton G. (1988). *Thematic Origins of Scientific Thought. Kepler to Einstein*, revised edn. Cambridge, MA: Harvard University Press.

Holton G. (1993). More on Mach and Einstein. In *Science and Anti-Science*. Cambridge, MA: Harvard University Press, 56–73.

Holton G. (1998a). Thematic presuppositions and the direction of scientific advance. In *The Advancement of Science and its Burdens*, 2 edn. Cambridge, MA: Harvard University Press, 3–27.

Holton G. (1998b). Einstein's model for constructing a scientific theory. In *The Advancement of Science and its Burdens*, 2 edn. Cambridge, MA: Harvard University Press, 28–56.

Holton G. (2005). Werner Heisenberg and Albert Einstein. In *Victory and Vexation in Science. Einstein, Bohr, Heisenberg and Others*. Cambridge, MA: Harvard University Press, 26–35.

Home D., Whitaker A. (2007). *Einstein's Struggles with Quantum Theory. A Reappraisal.* New York: Springer.

Howard D. A. (1984). Realism and conventionalism in Einstein's philosophy of science: The Einstein–Schlick correspondence. *Philosophia Naturalis* **21**, 616–629.

Howard D. A. (1985). Einstein on locality and separability. *Studies in History and Philosophy of Science* **16**, 171–201.

Howard D. A. (1990a). Einstein and Duhem. *Synthese* **83**, 363–384.

Howard D. A. (1990b). "Nicht sein kann was nicht sein darf," or the prehistory of EPR, 1909–1935: Einstein's early worries about the quantum mechanics of composite systems. In *Sixty-two Years of Uncertainty. Historical, Philosophical and Physical Inquiries into the Foundations of Quantum Mechanics*, A. I. Miller, ed. New York: Plenum, 61–111.

Howard D. A. (2004). Einstein's philosophy of science. In *The Stanford Encyclopedia of Philosophy*. http://plato.stanford.edu/entries/einstein-philscience/. Downloaded on 5 August 2007.

Howard D. A. (2005). Albert Einstein as a philosopher of science. *Physics Today* **58/12**, 34–40.

Howard D. A. (Forthcoming). Einstein and the development of twentieth-century philosophy of science. In *The Cambridge Companion to Einstein*, M. Janssen and C. Lehner, eds. Cambridge: Cambridge University Press.

Infeld L., van der Waerden B. L. (1933). Die Wellengleichung des Elektrons in der allgemeinen Relativitätstheorie. *Sitzungsberichte der Preußischen Akademie der Wissenschaften*, 380–401.

Isaacson W. (2007). *Einstein: His Life and Universe*. New York: Simon and Schuster.

Jammer M. (1989). *The Conceptual Development of Quantum Mechanics*, 2 edn. Los Angeles, CA: Tomash.

Janssen M. (1999). Rotation as the nemesis of Einstein's Entwurf theory. In *The Expanding Worlds of General Relativity*, H. Goenner, J. Renn, J. Ritter and T. Sauer, eds. Boston, MA: Birkhäuser, 127–157.

Janssen M. (2005). Of pots and holes: Einstein's bumpy road to general relativity. *Annalen der Physik* **14**, Supplement, 58–85.

Janssen M. (2007). What did Einstein know and when did he know it? A Besso memo dated August 1913. In *The Genesis of General Relativity*, M. Janssen, J. D. Norton, J. Renn, T. Sauer and J. Stachel, eds. Vol. 2, Einstein's Zurich Notebook: Commentary and Essays. Dordrecht: Springer, 785–837.

Janssen M., Renn J. (2007). Untying the knot: How Einstein found his way back to field equations discarded in the Zurich notebook. In *The Genesis of General Relativity*, M. Janssen, J. D. Norton, J. Renn, T. Sauer and J. Stachel, eds. Vol. 2, Einstein's Zurich Notebook: Commentary and Essays. Dordrecht: Springer, 839–925.

Janssen M., Schulmann R., Illy J., Lehner C., Buchwald D. K., Eds. (2002). *The Collected Papers of Albert Einstein*. Vol. 7, Writings 1918–1921. Princeton, NJ: Princeton University Press.

Janssen M., Norton J. D., Renn J., Sauer T., Stachel J., Eds. (2007a). *The Genesis of General Relativity*. Vol. 1, Einstein's Zurich Notebook: Introduction and Source. Dordrecht: Springer.

Janssen M., Norton J. D., Renn J., Sauer T., Stachel J., Eds. (2007b). *The Genesis of General Relativity*. Vol. 2, Einstein's Zurich Notebook: Commentary and Essays. Dordrecht: Springer.

Janssen M., Renn J., Sauer T., Norton J. D., Stachel J. (2007c). A commentary on the notes on gravity in the Zurich notebook. In *The Genesis of General Relativity*, M. Janssen, J. D. Norton, J. Renn, T. Sauer and J. Stachel, eds. Vol. 2, Einstein's Zurich Notebook: Commentary and Essays. Dordrecht: Springer, 489–714.

Joas C., Lehner C. (2009). The classical roots of wave mechanics: Schrödinger's transformation of the optical-mechanical analogy. *Studies in History and Philosophy of Modern Physics* **40**, 338–351.

Jost R: (1980). Comment on "Einstein on particles, fields and the quantum theory." In *Some Strangeness in the Proportion. A Centennial Symposium to Celebrate the Achievements of Albert Einstein*, H. Woolf, ed. Reading, MA: Addison-Wesley, 252–265.

Kaiser D. (2002). Cold war requisitions, scientific manpower and the production of American physicists after World War II. *Historical Studies in the Physical and Biological Sciences* **33**, 131–159.

Kaluza T. (1921). Zum Unitätsproblem der Physik. *Sitzungsberichte der Preußischen Akademie der Wissenschaften*, 966–972.

Kennefick D. (2005). Einstein and the problem of motion: a small clue. In *The Universe of General Relativity*, A. J. Kox and J. Eisenstaedt, eds. Boston, MA: Birkhäuser, 109–124.

Kennefick D. (2007). *Traveling at the Speed of Thought. Einstein and the Quest for Gravitational Waves*. Princeton, NJ: Princeton University Press.

Kichenassamy S. (1992). Dirac equations in curved spacetime. In *Studies in the History of General Relativity*, J. Eisenstaedt and A. J. Kox, eds. Boston, MA: Birkhäuser, 383–392.

Kilmister C. (1994). *Eddington's Search for a Fundamental Theory*. Cambridge: Cambridge University Press.

Klein M. J. (1964). Einstein and the wave-particle duality. *The Natural Philosopher* **3**, 3–49.

Klein M. J. (1970). *Paul Ehrenfest*. Vol. 1, The Making of a Theoretical Physicist. Amsterdam: North-Holland.

Klein M. J., Kox A. J., Schulmann R., Eds. (1993). *The Collected Papers of Albert Einstein*. Vol. 5, Correspondence 1912–1914. Princeton, NJ: Princeton University Press.

Klein M. J., Kox A. J., Renn J., Schulmann R., Eds. (1995). *The Collected Papers of Albert Einstein*. Vol. 4, Writings 1912–1914. Princeton, NJ: Princeton University Press.

Klein O. (1926a). The atomicity of electricity as a quantum theory law. *Nature* **118**, 516.

Klein O. (1926b). Quantentheorie und fünfdimensionale Relativitätstheorie. *Zeitschrift für Physik* **37**, 895–906.

Kox A. J. (1988). Hendrik Antoon Lorentz, the ether, and the general theory of relativity. *Archive for History of Exact Sciences* **38**, 67–78.

Kox A. J., Ed. (2008). *The Scientific Correspondence of H. A. Lorentz*. Vol. 1. New York: Springer.

Kox A. J., Klein M. J., Schulmann R., Eds. (1996). *The Collected Papers of Albert Einstein*. Vol. 6, Writings 1915–1917. Princeton, NJ: Princeton University Press.

Kragh H. (1982). Erwin Schrödinger and the wave equation: The crucial phase. *Centaurus* **26**, 154–197.

Kuhn T. S. (1978). *Black-body Theory and the Quantum Discontinuity 1894–1912*. Oxford: Clarendon Press.

Langer R. M. (1933). The new quantum mechanics. *Science* **77**, 6a.

Lenzen V. (1949). Einstein's theory of knowledge. In *Albert Einstein: Philosopher-Scientist*, P. A. Schilpp, ed. La Salle, IL: Open Court, 355–384.

Levenson T. (2003). *Einstein in Berlin*. New York: Bantam.

Lieb E. H., Simon B., Wightman A. S. (1976). *Studies in Mathematical Physics: Essays in Honour of V. Bargmann*. Princeton, NJ: Princeton University Press.

Maas A. (2007). Einstein as engineer: The case of the little machine. *Physics in Perspective* **9**, 305–328.

Mach E. (1912). *Die Mechanik in ihrer Entwicklung historisch-kritisch dargestellt*, 7 edn. Leipzig: F. A. Brockhaus.

Mach E. (1921). *Die Prinzipien der physikalischen Optik. Historisch und erkenntnispsychologisch entwickelt*. Leipzig: J. A. Barth.

Massimi M. (2005). *Pauli's Exclusion Principle. The Origin and Validation of a Scientific Principle*. Cambridge: Cambridge University Press.

McCormmach R. (1970). H. A. Lorentz and the electromagnetic view of nature. *Isis* **61**, 459–497.

Mehra J., Rechenberg H. (1982). *The Historical Development of Quantum Mechanics*. Vol. 2, The Discovery of Quantum Mechanics. Berlin: Springer.

Mie G. (1914a). Bemerkungen zu der Einsteinschen Gravitationstheorie. *Physikalische Zeitschrift* **15**, 115–122.

Mie G. (1914b). Bemerkungen zu der Einsteinschen Gravitationstheorie. II. *Physikalische Zeitschrift* **15**, 169–176.

Miller A. I. (1992). Albert Einstein's 1907 Jahrbuch paper: The first step from SRT to GRT. In *Studies in the History of General Relativity*, J. Eisenstaedt and A. J. Kox, eds. Boston, MA: Birkhäuser, 319–335.

Millikan R. A. (1949). Albert Einstein on his seventieth birthday. *Reviews of Modern Physics* **21**, 343–345.

Morrison M. (2000). *Unifying Scientific Theories*. Cambridge: Cambridge University Press.

Morus I. R. (2005). *When Physics Became King*. Chicago, IL: University of Chicago Press.

Moszkowski A. (1919). Die Sonne bracht' es an den Tag! *Berliner Tageblatt*, 8 October, evening edition.

Muller F. A. (1997). The equivalence myth of quantum mechanics, Parts I and II. *Studies in History and Philosophy of Modern Physics* **28**, 35–61, 219–247.

Neffe J. (2007). *Einstein. A Biography*. New York: Farrar, Straus and Giroux.

Nordström G. (1912). Relativitätsprinzip und Gravitation. *Physikalische Zeitschrift* **13**, 1126–1129.

Norton J. D. (1984). How Einstein found his field equations, 1912–1915. *Historical Studies in the Physical Sciences* **31**, 253–316.

Norton J. D. (1985). What was Einstein's principle of equivalence? *Studies in History and Philosophy of Science* **16**, 203–246.

Norton J. D. (1992). Einstein, Nordström and the early demise of scalar, Lorentz-covariant theories of gravitation. *Archive for History of Exact Sciences* **45**, 17–94.

Norton J. D. (2000). "Nature is the realisation of the simplest conceivable mathematical ideas": Einstein and the canon of mathematical simplicity. *Studies in History and Philosophy of Modern Physics* **31**, 135–170.

Norton J. D. (2007). What was Einstein's "Fateful Prejudice"? In *The Genesis of General Relativity*, M. Janssen, J. D. Norton, J. Renn, T. Sauer and J. Stachel, eds. Vol. 2, Einstein's Zurich Notebook: Commentary and Essays. Dordrecht: Springer, 715–783.

O'Raifeartaigh L. (1997). *The Dawning of Gauge Theory*. Princeton, NJ: Princeton University Press.

O'Raifeartaigh L., Straumann N. (2000). Gauge theory: Historical origins and some modern developments. *Reviews of Modern Physics* **72**, 1–23.

Pais A. (1980). Einstein on particles, fields and the quantum theory. In *Some Strangeness in the Proportion. A Centennial Symposium to Celebrate the Achievements of Albert Einstein*, H. Woolf, ed. Reading, MA: Addison-Wesley, 197–251.

Pais A. (1982). *'Subtle is the Lord...' The Science and the Life of Albert Einstein*. Oxford: Oxford University Press.

Pais A. (2000). *The Genius of Science: A Portrait Gallery of Twentieth Century Physicists*. Oxford: Oxford University Press.

Pauli W. (1921). Relativitätstheorie. In *Encyklopädie der mathematischen Wissenschaften, mit Einschluss ihrer Anwendungen*, A. Sommerfeld, ed. Vol. 5, Physik, part 2. Leipzig: Teubner, 539–775.

Pauli W. (1925). Über den Zusammenhang des Abschlusses der Elektronengruppen im Atom mit der Komplexstruktur der Spektren. *Zeitschrift für Physik* **31**, 765–783.

Pauli W. (1933a). Einige die Quantenmechanik betreffenden Erkundigungsfragen. *Zeitschrift für Physik* **80**, 573–586.

Pauli W. (1933b). Über die Formulierung der Naturgesetze mit fünf homogenen Koordinaten. Teil I. Klassische Theorie. Teil II. Die Diracschen Gleichungen für die Materiewellen. *Annalen der Physik* **18**, 305–336, 337–372.

Pauli W. (1958). *Theory of Relativity*. London: Pergamon.

Pauli W., Solomon J. (1932). La théorie unitaire d'Einstein–Mayer et les équations de Dirac. Parts I and II. *Journal de Physique et le Radium* **3**, 452–463, 582–589.

Penrose R., Rindler W. (1984). *Spinors and Spacetime*. Vol. I, Two-Spinor Calculus and Relativistic Fields. Cambridge: Cambridge University Press.

Perovic S. (2008). Why were matrix mechanics and wave mechanics considered equivalent? *Studies in History and Philosophy of Modern Physics* **39**, 444–461.

Popper K. (1935). *Logik der Forschung. Zur Erkenntnistheorie der modernen Naturwissenschaft*. Vienna: Springer.

Popper K. (1959). *The Logic of Scientific Discovery*. London: Hutchinson.

Renn J. (2005a). Before the Riemann tensor: the emergence of Einstein's double strategy. In *The Universe of General Relativity*, A. J. Kox and J. Eisenstaedt, eds. Boston, MA: Birkhäuser, 53–65.

Renn J. (2005b). Einstein's invention of Brownian motion. *Annalen der Physik* **14**, Supplement, 23–37.

Renn J. (2006). *Auf den Schultern von Riesen und Zwergen. Einsteins unvollendete Revolution*. Weinheim: Wiley-VCH.

Renn J. (2007). The summit almost scaled: Max Abraham as a pioneer of a relativistic theory of gravitation. In *The Genesis of General Relativity*, J. Renn and M. Schemmel, eds. Vol. 3, Gravitation in the Twilight of Classical Physics: Between Mechanics, Field Theory, and Astronomy. Dordrecht: Springer, 305–330.

Renn J., Sauer T. (1999). Heuristics and mathematical representation in Einstein's search for a gravitational field equation. In *The Expanding Worlds of General Relativity*, H. Goenner, J. Renn, J. Ritter and T. Sauer, eds. Boston, MA: Birkhäuser, 87–125.

Renn J., Sauer T. (2007). Pathways out of classical physics: Einstein's double strategy in his search for the gravitational field equation. In *The Genesis of General Relativity*, M. Janssen, J. D. Norton, J. Renn, T. Sauer and J. Stachel, eds. Vol. 1, Einstein's Zurich Notebook: Introduction and Source. Dordrecht: Springer, 113–312.

Renn J., Schemmel M., Eds. (2007a). *The Genesis of General Relativity*. Vol. 3, Gravitation in the Twilight of Classical Physics: Between Mechanics, Field Theory, and Astronomy. Dordrecht: Springer.

Renn J., Schemmel M., Eds. (2007b). *The Genesis of General Relativity*. Vol. 4, Gravitation in the Twilight of Classical Physics: The Promise of Mathematics. Dordrecht: Springer.

Renn J., Sauer T., Stachel J. (1997). The origin of gravitational lensing: A postscript to Einstein's 1936 Science paper. *Science* **275**, 184–186.

Roberston H. P., Noonan T. (1968). *Relativity and Cosmology*. Philadelphia, PA: Saunders.

Rowe D. E. (2002). Einstein's gravitational field equations and the Bianchi identities. *The Mathematical Intelligencer* **24**, 4, 57–70.

Rowe D. E. (2006). Einstein's allies and enemies: Debating relativity in Germany, 1916–1920. In *Interactions: Physics, Mathematics and Philosophy, 1860–1930*, V. F. Hendricks, K. F. Jørgensen, J. Lützen and S. A. Pedersen, eds. Dordrecht: Springer, 231–280.

Rowe D. E., Schulmann R., Eds. (2007). *Einstein on Politics. His Private Thoughts and Public Stands on Nationalism, Zionism, War, Peace and the Bomb*. Princeton, NJ: Princeton University Press.

Rüchardt E. (1926). E. Rupp. Interferenzuntersuchungen an Kanalstrahlen. *Physikalische Berichte* **7**, 1523–1524.

Rupp E. (1926a). Interferenzuntersuchungen an Kanalstrahlen. *Annalen der Physik* **79**, 1–34.

Rupp E. (1926b). Über die Interferenzeigenschaften des Kanalstrahllichtes. *Sitzungsberichte der Preußischen Akademie der Wissenschaften*, 341–351.

Ryckman T. (2005). *The Reign of Relativity. Philosophy in Physics 1915–1925*. Oxford: Oxford University Press.

Rynasiewicz R. (1994). The lessons of the hole argument. *British Journal for the Philosophy of Science* **45**, 407–436.

Sauer T. (1999). The relativity of discovery: Hilbert's first note on the foundations of physics. *Archive for History of Exact Sciences* **53**, 529–575.

Sauer T. (2005). Einstein equations and Hilbert action: What is missing on page 8 of the proofs for Hilbert's first communicaton on the foundations of physics? *Archive for History of Exact Sciences* **59**, 577–590.

Sauer T. (2006). Field equations in teleparallel space–time: Einstein's Fernparallelismus approach toward unified field theory. *Historia Mathematica* **33**, 399–439.

Sauer T. (2007a). Einstein and the early theory of superconductivity, 1919–1922. *Archive for History of Exact Sciences* **61**, 159–211.

Sauer T. (2007b). The Einstein–Varičak correspondence on relativistic rigid rotation. In *Proceedings of the Eleventh Marcel Grossmann Meeting on General Relativity*, H. Kleinert and R. T. Jantzen, eds. Singapore: World Scientific, 2453–2455.

Sauer T. (Forthcoming). Einstein's unified field theory program. In *The Cambridge Companion to Einstein*, M. Janssen and C. Lehner, eds. Cambridge: Cambridge University Press.

Scherrer W. (1934). Quaternionen und Semivektoren. *Commentarii Mathematici Helvetici* **7**, 141–149.

Schlick M. (1915). Die philosophische Bedeutung des Relativitätsprinzips. *Zeitschrift für Philosophie und philosophische Kritik* **159**, 129–175.

Schlick M. (1917). *Raum und Zeit in der gegenwärtigen Physik*. Berlin: Springer.

Schlick M. (1979). The philosophical significance of the principle of relativity. In *Philosophical Papers*, H. L. Mulder and B. F. B. van de Velde-Schlick, eds. Vol. I, 1909–1922. Dordrecht: Reidel, 153–189.

Scholz E., Ed. (2001). *Hermann Weyl's Raum-Zeit-Materie and a General Introduction to his Scientific Work*. Basel: Birkhäuser.

Scholz E. (2006). The changing concept of matter in H. Weyl's thought, 1918–1930. In *Interactions: Physics, Mathematics and Philosophy, 1860–1930*, V. F. Hendricks, K. F. Jørgensen, J. Lützen and S. A. Pedersen, eds. Dordrecht: Springer, 281–306.

Schönbeck C. (2000). Albert Einstein und Philipp Lenard. Antipoden in Physik und Zeitgeschichte. *Schriften der mathematisch–naturwissenschaftliche Klasse der Heidelberger Akademie der Wissenschaften* **8**, 1–42.

Schouten J. A. (1933). Zur generellen Feldtheorie. Semivektoren und Spinraum. *Zeitschrift für Physik* **84**, 92–111.

Schulmann R., Kox A. J., Janssen M., Illy J., Eds. (1998). *The Collected Papers of Albert Einstein*. Vol. 8, Correspondence 1914–1918. Princeton, NJ: Princeton University Press.

Schwarzschild K. (1916). Über das Gravitationsfeld eines Massenpunktes nach der Einsteinschen Theorie. *Sitzungsberichte der Königlich Preußischen Akademie der Wissenschaften*, 189–196.

Schweber S. S. (1986). The empiricist temper regnant: Theoretical physics in the United States, 1920–1950. *Historical Studies in the Physical and Biological Sciences* **17**, 55–98.

Schweber S. S. (2008). *Einstein and Oppenheimer. The Meaning of Genius*. Cambridge, MA: Harvard University Press.

Seth S. (2004). Quantum theory and the electromagnetic world view. *Historical Studies in the Physical and Biological Sciences* **35**, 67–93.

Seth S. (2007). Crisis and the construction of modern theoretical physics. *British Journal for the History of Science* **40**, 25–51.

Seth S. (2008). Crafting the quantum: Arnold Sommerfeld and the older quantum theory. *Studies in History and Philosophy of Science* **39**, 335–348.

Shankland R. S. (1964). Michelson–Morley experiment. *American Journal of Physics* **32**, 16–35.

Smeenk C., Martin C. (2007). Mie's theories of matter and gravitation. In *The Genesis of General Relativity*, J. Renn and M. Schemmel, eds. Vol. 4, Gravitation in the Twilight of Classical Physics: The Promise of Mathematics. Dordrecht: Springer, 623–632.

Smolin L. (2006). *The Trouble with Physics. The Rise of String Theory, the Fall of a Science, and What Comes Next*. Boston, MA: Houghton Mifflin.

Sommerfeld A. (1916). Zur Quantentheorie der Spektrallinien. *Annalen der Physik* **51**, 1–94; 125–167.

Sommerfeld A. (1920). Allgemeine spektroskopische Gesetze, insbesondere ein magnetooptischer Zerlegungssatz. *Annalen der Physik* **63**, 221–263.

Sorkin R. (1983). Kaluza–Klein monopole. *Physical Review Letters* **51**, 87–90.

Stachel J. (1986). Einstein and the quantum: Fifty years of struggle. In *From Quarks to Quasars: Philosophical Problems in Modern Physics*, R. G. Colodny, ed. Pittsburgh, PA: University of Pittsburgh Press, 349–385.

Stachel J. (1989a). The rigidly rotating disk as the "missing link" in the history of general relativity. In *Einstein and the History of General Relativity*, D. A. Howard and J. Stachel, eds. Boston, MA: Birkhäuser, 48–62.

Stachel J. (1989b). Einstein's search for general covariance, 1912–1915. In *Einstein and the History of General Relativity*, D. A. Howard and J. Stachel, eds. Boston, MA: Birkhäuser, 63–100.

Stachel J. (1991). Einstein and quantum mechanics. In *Conceptual Problems of Quantum Gravity*, A. Ashtekar and J. Stachel, eds. Boston, MA: Birkhäuser, 13–42.

Stachel J. (1993). The other Einstein: Einstein contra field theory. *Science in Context* **6**, 275–290.

Stachel J. (2002). *Einstein from 'B' to 'Z'*. Boston, MA: Birkhäuser.

Stachel J. (2007). The first two acts. In *The Genesis of General Relativity*, M. Janssen, J. D. Norton, J. Renn, T. Sauer and J. Stachel, eds. Vol. 1, Einstein's Zurich Notebook: Introduction and Source. Dordrecht: Springer, 81–111.

Staley R. (2005). On the co-creation of classical and modern physics. *Isis* **96**, 530–558.

Stern F. (1999). *Einstein's German World*. Princeton, NJ: Princeton University Press.

Straub H. (1930). Über die Kohärenzlänge des von Kanalstrahlen emittierten Leuchtens. *Annalen der Physik* **5**, 644–656.

Study E. (1913). *Die realistische Weltansicht und die Lehre vom Raume. Geometrie, Anschauung und Erfahrung*. Braunschweig: Vieweg.

Tauschinsky A., van Dongen J. (2008). Over lichtemissie: Albert Einstein en de vroege geschiedenis van de Nederlandse Natuurkundige Vereniging. *Nederlands Tijdschrift voor Natuurkunde* **74**, 138–141.

ter Haar D. (1967). *The Old Quantum Theory*. Oxford: Pergamon.

Thiry Y. (1948). Les équations de la théorie unitaire de Kaluza. *Comptes Rendus de l'Académie des Sciences* **226**, 216–218.

Thomson G. P., Reid A. (1927). Diffraction of cathode rays by a thin film. *Nature* **119**, 890.

Tonnelat M. (1965). *Les Théories Unitaires de l'Électromagnétisme et de la Gravitation.* Paris: Gauthier-Villars.

Ullmo J. (1934). Quelques propriétés du groupe de Lorentz. Semi-vecteurs et spinors. *Journal de Physique et le Radium* **5**, 230–240.

Unna I. (2000). The genesis of physics at the Hebrew University of Jerusalem. *Physics in Perspective* **2**, 336–380.

van der Waerden B. L. (1929). Spinoranalyse. *Nachrichten von der Gesellschaft der Wissenschaften zu Göttingen*, 100–109.

van der Waerden B. L. (1932). *Die gruppentheoretische Methode in der Quantenmechanik.* Berlin: Springer.

van der Waerden B. L. (1960). Exclusion principle and spin. In *Theoretical Physics in the Twentieth Century. A Memorial Volume to Wolfgang Pauli*, M. Fierz and V. F. Weisskopf, eds. New York: Interscience, 199–244.

van der Waerden B. L., Ed. (1968). *Sources of Quantum Mechanics.* New York: Dover.

van Dongen J. (2002). Einstein and the Kaluza–Klein particle. *Studies in History and Philosophy of Modern Physics* **33**, 185–210.

van Dongen J. (2007a). Emil Rupp, Albert Einstein, and the canal ray experiments on wave-particle duality: Scientific fraud and theoretical bias. *Historical Studies in the Physical and Biological Sciences* **37**, Supplement, 73–120.

van Dongen J. (2007b). The interpretation of the Einstein–Rupp experiments and their influence on the history of quantum mechanics. *Historical Studies in the Physical and Biological Sciences* **37**, Supplement, 121–131.

van Dongen J. (2007c). Reactionaries and Einstein's fame: "German Scientists for the Preservation of Pure Science," relativity and the Bad Nauheim conference. *Physics in Perspective* **9**, 212–230.

van Dongen J. (2009). On the role of the Michelson–Morley experiment: Einstein in Chicago. *Archive for History of Exact Sciences* **63**, 655–663.

Vizgin V. P. (1994). *Unified Field Theories in the First Third of the Twentieth Century.* Boston, MA: Birkhäuser.

von Meyenn K., Ed. (1985). *Wolfgang Pauli. Wissenschaftlicher Briefwechsel mit Bohr, Einstein, Heisenberg u.a.* Vol. II, 1930–1939. Berlin: Springer.

von Neumann J. (1932). *Mathematische Grundlagen der Quantenmechanik.* Berlin: Springer.

Warwick A. (2003). *Masters of Theory: Cambridge and the Rise of Mathematical Physics.* Chicago, IL: University of Chicago Press.

Weinberg S. (1972). *Gravitation and Cosmology: Principles and Applications of the General Theory of Relativity.* New York: John Wiley & Sons.

Weinberg S. (1995). *The Quantum Theory of Fields.* Vol. I, Foundations. Cambridge: Cambridge University Press.

Wessels L. (1979). Schrödinger's route to wave mechanics. *Studies in History and Philosophy of Science* **10**, 311–340.

Weyl H. (1918). Gravitation und Elektrizität. *Sitzungsberichte der Königlich Preußischen Akademie der Wissenschaften*, 465–478, 478–480.

Weyl H. (1929). Elektron und Gravitation. I. *Zeitschrift für Physik* **56**, 330–352.

Weyl H. (1931a). Geometrie und Physik. *Die Naturwissenschaften* **19**, 49–58.

Weyl H. (1931b). *Gruppentheorie und Quantenphysik.* Leipzig: S. Hirzel.

Whitaker A. (1996). *Einstein, Bohr and the Quantum Dilemma.* Cambridge: Cambridge University Press.

Woit P. (2006). *Not Even Wrong. The Failure of String Theory and the Continuing Challenge to Unify the Laws of Physics.* London: Jonathan Cape.

Wolff S. L. (2000). Frederick Lindemanns Rolle bei der Emigration der aus Deutschland vertriebenen Physiker. In *Yearbook of the Research Centre for German and Austrian Exile Studies*, A. Grenville, ed. Vol. 2, German-Speaking Exiles in Great Britain. Amsterdam: Rodopi, 25–58.

Wolters G. (1984). Ernst Mach and the theory of relativity. *Philosophia Naturalis* **21**, 630–641.

Wünsch D. (2003). The fifth dimension: Theodor Kaluza's ground breaking idea. *Annalen der Physik* **12**, 519–542.

Wünsch D. (2005). Einstein, Kaluza, and the fifth dimension. In *The Universe of General Relativity*, A. J. Kox and J. Eisenstaedt, eds. Boston, MA: Birkhäuser, 277–302.

Index

Aarau, 76
Abraham, Max, 7, 9–10, 19, 22, 80
Academia Olympia, 51
Afanassjewa, Tatiana, 98
Albert Einstein Archives, 43
Ampère's molecular currents, 77–79
Anschaulichkeit, 171
anti-proton, 115, 116
anti-relativists, 65, 82
a priori judgment, 45–48
Atkinson, Robert d'E., 86, 88
atomists, 63
Avogadro's number, 76

Bad Nauheim, 65, 166, 186
Balmer series, 158
Bargmann, Valentin, 122, 128, 130, 143–145, 147–187
Bell, John S., 180
Beller, Mara, 59
Belousek, Darrin W., 181
Bergmann, Emmy, 139
Bergmann, Max, 139
Bergmann, Peter, 130, 139–146, 149, 151, 154–155, 187
Bernstein, Aaron, 60–61
Besso, Michele, 9, 10, 23, 25, 29, 39, 41, 92, 165, 169, 170
Bethe, Hans, 188
Bianchi identities, 16
bi-vectors, 146
BKS (Bohr–Kramers–Slater) theory, 87, 165, 169
black body spectrum, 91, 158, 164, 170
Black Mountain College, 141
Bohm, David, 2, 181–182
Bohr atom, 158–159, 164, 167
Bohr, Niels, 86, 99, 158–159, 161–163, 165, 168, 171, 173, 175, 187
Bohr–Sommerfeld quantization rules, 158
Bonfante, Julian H., 60
Bopp, Fritz, 188

Born interpretation, 172, 178
Born, Max, 47, 89, 163, 166–169, 172, 173, 178, 183, 187
Bose–Einstein theory, 164, 170
Bothe, Walther, 81, 86–87, 161
Breit, Gregory, 178
Bridgman, Percy W., 39, 188
Brownian motion, 76–77
Bucherer, Alfred, 80
Büchner, Ludwig, 60–61
Bucky, Gustav, 80

California Institute of Technology, 101, 102, 118, 125
canal rays, 83–86, 88
canonical quantum gravity, 155
Cartan, Élie, 67–68, 125
Cassidy, David, 146
Cato the Elder, 177
causality, 165, 171, 173, 174, 176, 178, 183
charge multiplier, Einstein's, 76–77
charge, quantum of, 120–121, 136, 144–145, 151–152, 154, 168, 178
Chicago, 90
Christoffel, Elwin, 28
classical physics, introduction of designation, 158
Cold War, 188
Columbia University, 40
compactification, of fifth dimension, 135, 136, 138, 142, 144, 151, 152, 154, 185
completeness, 174, 176, 179, 180, 182
configuration space, 170, 172, 176, 177
constructive theories, 49–51
conventionalism, 48
coordinate conditions
    harmonic, 16
    vs. coordinate restrictions, 16
Copenhagen interpretation, 172, 178, 187
Copernicus, Nicolaus, 34
core model, of atom, 161
correspondence principle, 168
cylinder condition, 133, 135, 139

208